DESIGNING OF ANALOG AND DIGITAL CONTROL SYSTEMS

J. L. MIN
Product Manager Industrial Projects
The Netherlands

and

J. J. SCHRAGE
Lecturer in Control Engineering
The Netherlands

Translators:
E. v. KAMPEN
J. L. MIN

ELLIS HORWOOD LIMITED
Publishers · Chichester

Halsted Press: a division of
JOHN WILEY & SONS
New York · Chichester · Brisbane · Toronto

This English edition first published in 1988 by
ELLIS HORWOOD LIMITED
Market Cross House, Cooper Street,
Chichester, West Sussex, PO19 1EB, England
The publisher's colophon is reproduced from James Gillison's drawing of the ancient Market Cross, Chichester.

Distributors:
Australia and New Zealand:
JACARANDA WILEY LIMITED
GPO Box 859, Brisbane, Queensland 4001, Australia

Canada:
JOHN WILEY & SONS CANADA LIMITED
22 Worcester Road, Rexdale, Ontario, Canada

Europe and Africa:
JOHN WILEY & SONS LIMITED
Baffins Lane, Chichester, West Sussex, England

North and South America and the rest of the world:
Halsted Press: a division of
JOHN WILEY & SONS
605 Third Avenue, New York, NY 10158, USA

This English edition is translated from the original Dutch edition *Ontwerpen van analogue en digitale regelsystemen,* published in 1982 by Nijgh & Van Ditmar Educatief, The Hague, The Netherlands © the copyright holders.

© **1988 English Edition, Ellis Horwood Limited**

British Library Cataloguing in Publication Data
Min, J. L., *1949–*
Designing analog and digital control systems.
1. Control systems. Applications of computer systems
I. Title   II. Schrage, J. J., *1938–*
III. Ontwerpen van analogue en digitale regelsystemen. *English*
629.8′312

Library of Congress Card No. 88–30139

ISBN 0–7458–0460–8 (Ellis Horwood Limited)
ISBN 0–470–21253–5 (Halsted Press)

Phototypeset in Times by Ellis Horwood Limited
Printed in Great Britain by Hartnolls, Bodmin

COPYRIGHT NOTICE
All Rights Reserved. No part of this publication may be reproduced, stored in a retrieval system, or transmitted, in any form or by any means, electronic, mechanical, photocopying, recording or otherwise, without the permission of Ellis Horwood Limited, Market Cross House, Cooper Street, Chichester, West Sussex, England.

Table of contents

Preface 9
1 GENERAL INTRODUCTION 11
2 SYSTEM DESCRIPTION METHODS
    2.1 Introduction 13
    2.2 System description in the t-domain 13
    2.3 System description in the w-domain 20
    2.4 System description in the s-domain 25
    2.5 System description in the z-domain 28
    2.6 Problems 29
3 SYSTEM DESCRIPTION IN THE s-DOMAIN
    3.1 Introduction 32
    3.2 The poles and zeros presentation 33
    3.3 The relation between the pole/zero plot and responses in the t-domain .37
    3.4 The relation between the pole/zero representation and
        the transfer function 47
    3.5 Problems 52
4 SYSTEM ANALYSIS IN THE s-PLANE
    4.1 Introduction 54
    4.2 Stability criteria in the s-plane 54
    4.3 Lines of absolute and relative damping in the s-plane 58
    4.4 Non-minimum-phase systems 65
    4.5 Systems with time delay 67
    4.6 Problems 68
5 ROOT LOCI SYSTEMS WITHOUT TIME DELAY
    5.1 Introduction 70
    5.2 The influence of feedback on the pole/zero representation of a system .70
    5.3 Construction and calculation rules for root loci 75
    5.4 Root loci for negative values of K 98

	5.5 Computer aided calculation and plotting of root loci 100	
	5.6 Problems 102	

6 ROOT LOCI OF SYSTEMS WITH TIME DELAY
   6.1 Introduction 106
   6.2 Phase lines in the s-domain 107
   6.3 Root locus construction by means of phase lines 109
   6.4 Approximation methods of a time delay 115
   6.5 Problems 117

7 DESIGNING CONTROL LED SYSTEMS IN THE s-DOMAIN
   7.1 Introduction 119
   7.2 Design criteria in the s-domain 120
   7.3 Designing a proportionally controlled system 122
   7.4 The addition of a derivative control action 126
   7.5 The addition of an integral control action 131
   7.6 Root loci in fluctuation of a time constant 134
   7.7 Computer aided design of a derivative control action 141
   7.8 Designing a derivative control action by means of the phase condition
       of the root locus 144
   7.9 Problems 150

8 THREE CASE STUDIES OF ANALOG CONTROL SYSTEMS
   8.1 Introduction 152
   8.2 A DC position servo system 152
   8.3 A phase lock loop 157
   8.4 The 'gravity raiser' 160

9 DIGITAL COMPUTER APPLICATION IN PROCESS CONTROL
   9.1 Introduction 166
   9.2 Off-line computer applications 171
   9.3 On-line applications 172
   9.4 In-line applications 173
   9.5 Problems 175

10 DIRECT DIGITAL CONTROL
   10.1 Introduction 177
   10.2 The control loop 177
   10.3 Interfaces 180
   10.4 Signal sampling 185
   10.5 Signal processing in the CPU 185
   10.6 Signal reconstruction 187
   10.7 Qualitative consideration of sampling and reconstruction 192
   10.8 Problems 196

11 MATHEMATICAL DESCRIPTION OF SAMPLED DATA SYSTEMS
   11.1 Introduction 198
   11.2 Description of a sampled data system by means of an impulse sequence 198

11.3 The Laplace transform of sampled data 202
  11.4 Fourier series of sampled data 204
  11.5 The z-transform 210
  11.6 The table of z-transforms and some properties of the z-transform. . . 213
  11.7 Inverse z-transform 215
  11.8 Problems 218

12 BLOCK DIAGRAMS OF SAMPLED DATA SYSTEMS
  12.1 Introduction 221
  12.2 Transfer functions of discrete elements and systems 222
  12.3 Transfer functions of sampled data systems 223
  12.4 Applications 232
  12.5 Problems 238

13 ANALYSIS AND SYNTHESIS OF SAMPLED DATA SYSTEMS IN THE z-DOMAIN
  13.1 Introduction 240
  13.2 Stability analysis in the z-domain 240
  13.3 Design criteria in the z-domain 247
  13.4 Proportional control design of a sampled data first order system. . . . 251
  13.5 Proportional control design of a sampled data second order system. . 255
  13.6 Problems 258

14 DIGITAL NETWORKS
  14.1 Introduction 261
  14.2 Some programming methods 261
  14.3 Control algorithms 267
  14.4 Examples of control algorithms 274
  14.5 Implementation of control algorithms 276
  14.6 Problems 278

15 PRACTICAL RULES FOR DIRECT DIGITAL CONTROL (DDC)
  15.1 Introduction 282
  15.2 Choosing a sampling frequency 282
  15.3 Adjustment of K_R, τ_i, and τ_d 284
  15.4 Digital process control design in the Bode diagram 288
  15.5 Comparison of analog and digital process control 289
  15.6 Problems 293

16 TIME-OPTIMAL CONTROL SYSTEMS
  16.1 Introduction 295
  16.2 The dead-beat response method 295
  16.3 Applications 299
  16.4 Appreciation of dead-beat control systems 305
  16.5 Problems 306

17 DIGITAL FILTERS
  17.1 Introduction 308

17.2 Filtering by increasing the sampling frequency 309
17.3 Signal filtering before sampling. 310
17.4 Signal filtering after sampling. 311
17.5 Problems 316

18 CASE STUDIES OF DIGITAL CONTROL SYSTEMS
18.1 Introduction 318
18.2 Phase lock loop with a sample and hold device 318
18.3 Temperature control with a digital PI controller 320
18.4 Set-up and operation of a digital PID controller 324
18.5 Application of a moving average filter 327

Answers 329

References and bibliography 347

Index 348

Preface

This book deals with the analysis and synthesis of continuous and data-sampled control systems with aid of the root-locus theory. The book is meant for Technical Colleges, but the authors hope that application and industrial courses will also make use of it.

In the first part of the book description methods of continuous system by means of a pole-zero-notation are introduced and the displacement of the poles under the influence of the fluctuation of a parameter is discussed. The properties of feedback systems in the time-domain are related on the location of the s-domain. The resulting design criteria are applied to the design of the P-, I- and D-actions in the s-domain.

In the second part of the book the characteristic properties of sampled-data systems are discussed qualitatively. Then handling treatment in the z-domain corresponds with the used root-locus method used in the first part.

The design criteria of the first part are recorded in the z-domain as well and the design procedures are discussed. Also the programming of the matching control algorithms is given schematically. The design of time-optimal control systems is explained and the results are evaluated. Some realizations of continuous as well as sampled-data system are discussed in depth. Each chapter makes ample use of detailed examples and ends with several exercises. The brief solutions to the exercises, as well as references, are included at the end of the book. The several methods for analysis and design are compared regularly with simulation results. In many cases these results have been realized with aid of the simulation program TUTSIM, developed by the University of Technology in Twente (The Netherlands).

The control theory discussed in this book has been given increasing attention, and with the rise of the microcomputer there has been a greater number of practical implementations. The authors would be pleased to learn of the experiences of colleagues working in the same field; their remarks and contributions will be taken into account in the next edition. We are grateful to Mr Michielsen for his cover design.

Preface to the second edition

Owing to the great interest in the first edition, this second edition is published shortly after the first one. We have had the opportunity to correct some mistakes that have been pointed out and to digest the various remarks directly.

From a didactical point of view some modifications are made in the sequence of the lessons. A new paragraph is added that deals with the course of the root-locus in the case of a positive feedback system.

We do hope that this second edition (like the first one) will be constructively criticized.

Preface to the third edition

In this third edition some remaining mistakes are corrected. Chapter 14 has been modified in such a way that the comprehensibility is increased, while more attention has also been paid to the implementation control algorithms. A practical realization of a moving average filter is added to Chapter 18.

Remarks leading to any improvement of this edition will be welcomed.

J. L. Min
J. J. Schrage

1
General introduction

The application of digital computers in control engineering has increased enormously during the last few years. This development was predictable but has been speeded up by economic developments in the computer market. The price-performance ratio has increased astonishingly in a very short time, especially because of the development of the microprocessor on one or a few semiconductor chips. Also, for less complex processing systems the application of digital computers is now profitable.

Some ten years ago the use of digital computers became necessary in space technology and, in complex — usually chemical — processes. The complex technical problems that occur in controlling rockets and chemical processes — these are mostly multivariable systems — can only be solved through an integral approach by means of a computer. The high cost of this apparatus had to be accepted. At the moment, and this applies especially to hardware, the price of computer control systems is ever more competitive with more conventional solutions like the classical PID-control. Moreover, the application of computers often yields additional advantages that are becoming more and more important. One may think of quality control and data logging.

There are two main areas in which we use digital computers in process engineering:

- The use for analysis and synthesis of systems to be controlled or controlling respectively.
- The use as part of a control.

As to the use in these two categories we need not make many divergent demands upon the digital computer itself, although the second category requires a high calculating speed. The configuration, however, must be adapted to the objective. Thus, if we use a digital computer for analysis and synthesis objectives, the man–computer communication must be carefully considered. If the computer is used in a control loop, the communication between process and computer will make signal adaptation necessary. In addition, as part of a control loop, a digital computer influences the process to be controlled in an essentially different way from a conventional (analog) controller, because signal processing cannot take place continuously. This results in extra knowledge being necessary for the adjustment of the control parameters.

The first part of this book mostly concerns the design of controlled systems in which we apply the pole/zero notation as a method of description. This approach may enable us to obtain usable results even without a computer. Understanding of control engineering, and especially of the relationship between the methods of description in the t-domain and the s-domain, may be much helped by familiarity with pole/zero notation. Moreover, this method of description is more and more being applied in electronics and telecommunication.

The second part of this book concerns the control systems in which the computer is integrated in the control loop. The use of the computer as a digital PID control and as a digital filter is explored, description in the z-domain being used. In this domain we may determine properties of systems in which the computer is part of one or more control loops. The methods and rules to be applied rest largely on the pole-zero theory used in the first part. The extra components, such as analog/digital and digital/analog converters, which are necessary for the integration of the computer in the control loop, are discussed. The results of sampling and reconstruction are specially considered; not only numerical solutions will be sought after but also a qualitative assessment, which often leads to greater insight.

2
System description methods

2.1 INTRODUCTION

In control engineering there are a number of ways to describe the relevant properties of a process or system. It often seems as if every method has its own supporters. This often creates communication problems for the supporters of different methods. Yet it is a fact that every new method has been developed because existing methods could not or could only with difficulty describe specific properties. Also the appearance of new mathematical methods can be a motive for applying newer methods of description in control engineering. The Laplace transform, first applied in mathematics, led to the description of continuous control systems in the so-called s-domain. The problems in describing discrete signals by means of the then known methods led to the creation of the z-transform, in which the system behaviour is laid down in the so-called z-domain. In this chapter the various methods will be illustrated briefly, especially taking into account those properties that will be used later in the book.

2.2 SYSTEM DESCRIPTION IN THE t-DOMAIN

In the time-domain, the behaviour of linear time independent systems with input $x(t)$ and output $y(t)$ is described by a differential equation of the form:

$$a_n \frac{d^n y(t)}{dt^n} + a_{n-1} \frac{d^{n-1} y(t)}{dt^{n-1}} + \ldots + a_1 \frac{dy(t)}{dt} + a_0 y(t) = \\ b_m \frac{d^m x(t)}{dt^m} + b_{m-1} \frac{d^{m-1} x(t)}{dt^{m-1}} + \ldots + b_1 \frac{dx(t)}{dt} + b_0 x(t) \ . \tag{2.1}$$

Always, m is at most equal to n: for a physically realizable system, however, n is always $>m$. The coefficients a_i and b_i are all real and time independent. The DC-gain or static gain of the system, that is, the relation between $y(t)$ and $x(t)$ if all transients are damped out, is represented by:

$$K = \frac{b_0}{a_0} \tag{2.2}$$

Example 2.1
In the electrical system of Fig. 2.1, at $t = 0$ the capacitor is uncharged, and equations

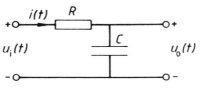

Fig. 2.1 — RC system.

(2.3), (2.4) apply:

$$u_i(t) = i(t) \cdot R + u_0(t) \tag{2.3}$$

$$u_0(t) = \frac{q(t)}{C} = \frac{1}{C} \int_0^t i(t) \, dt \tag{2.4}$$

From (2.3) and (2.4) we obtain the system differential equation:

$$RC \frac{du_o(t)}{dt} + u_0(t) = u_i(t) \tag{2.5}$$

This system is called a first order process or imperfect integrator with a dc gain of one and a time constant $\tau = RC$ seconds.

Example 2.2
The mechanical system of Fig. 2.2 is at rest, so that applies:

$$F_v = C_v(x(t) - y(t)) \tag{2.6}$$

$$F_v = F_d = C_d \frac{dy(t)}{dt} \tag{2.7}$$

From (2.6) and (2.7) we obtain the system differential equation:

Sec. 2.2] System description in the *t*-domain

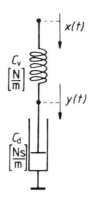

Fig. 2.2 — Spring–damper system.

$$\frac{C_d}{C_v} = \frac{dy(t)}{dt} + y(t) = x(t) \tag{2.8}$$

This system is also a first order system; the time constant is now $\tau = C_d/C_v$ seconds; the static gain is again 1.

From these examples it is clear that physically different processes can be described by differential equations of the same form.

If a system is described by the system differential equation, it is possible to calculate the output from a given input. Therefore the input is substituted in the differential equation, and this equation is solved by a standard procedure.

Example 2.3
A system with input $x(t)$ and output $y(t)$ is described by the differential equation:

$$2\frac{dy(t)}{dt} + y(t) = 4\,x(t) \tag{2.9}$$

At $t = 0$ a step-like signal with an amplitude of 10 is put on to the input, or:

$$x(t) = 10\,1(t) \tag{2.10}$$

with:

$$\begin{aligned}1(t) &= 0 \text{ for } t<0 \\ 1(t) &= 1 \text{ for } t>0\end{aligned} \tag{2.11}$$

For $t = 0$ this step function is not defined.

For $y(t)$:

$$2\frac{dy(t)}{dt} + y(t) = 40\ 1(t) \tag{2.12}$$

For this linear first order differential equation we may use this solution procedure:
First, the equation is reduced and then the roots of the characteristic equation are defined. In this way we find the complementary solution of the reduced differential equation. Next we try to find a particular solution of the complete differential equation by substituting the general form of the right part in this equation. From the sum of the complementary solution of the reduced equation and the particular solution of the complete equation we obtain the complete solution.
This procedure, applied to (2.12), gives:

- reduced differential equation: $2\frac{dy(t)}{dt} + y(t) = 0$
- characteristic equation: $2p + 1 = 0$
- root of the characteristic equation: $p = -\frac{1}{2}$.
- complementary solution of the reduced differential equation: $y(t) = C_1 e^{-\frac{1}{2}t}$
- try the particular solution: $y(t) = A$
- it gives: $A = 40\ 1(t)$
- the full complete solution will be

$$y(t) = (40 + C_1 e^{-\frac{1}{2}t})\ 1(t) \tag{2.13}$$

From the initial condition:

$$y(t) = 0 \text{ for } t \leqq 0 \tag{2.14}$$

it is possible to obtain $C_1 = -40$, which gives the output

$$y(t) = 40(1 - e^{-\frac{1}{2}t})\ 1(t) \tag{2.15}$$

For a quick evaluation of the system behaviour, the response of a (simple) system can be plotted on a known input signal. In this way we can also try to define the system's parameters, for example, time constants. For simple systems this mostly works out; for more complex systems it often doesn't.
Frequently used signals are step, ramp, pulse, and sinusoidal signals.

Example 2.4
A graph of the output calculated in Example 2.3, expression (2.15), is asked for. The response is the superposition — a property of linear systems — of two components,

namely $y_1(t) = 40 \cdot 1(t)$ and $y_2 = -40 \, e^{-\frac{1}{2}t} \, 1(t)$. These two components and their resultant are shown in Fig. 2.3.

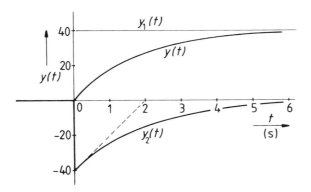

Fig. 2.3 — Step response of the first order system.

Example 2.5
As indicated in Fig. 2.4, a system gives the response $x(t)$ to a ramp input $y(t)$. We wish to determine the differential equation of the system.

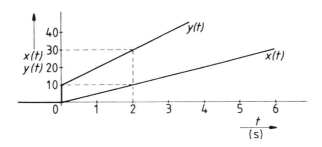

Fig. 2.4 — Response to a ramp test signal.

The step like change just after $t = 0$ indicates a differential in the system. The slope of $y(t)$ after $t = 0$ is a proportional contribution of $x(t)$ in $y(t)^-$.

$$y(t) = 2\frac{\mathrm{d}x(t)}{\mathrm{d}t} + 2x(t) \tag{2.16}$$

Every system can be imagined to be composed from one or more basic systems in cascade; we recognize the following basic systems:

1. *Constant factor*
   The system differential equation is:

   $$y(t) = Kx(t) \tag{2.17}$$

2. *Pure differentiator*
   With:

   $$y(t) = \tau_d \frac{dx(t)}{dt} \tag{2.18}$$

3. *Pure integrator*
   With:

   $$y(t) = \frac{1}{\tau_i} \int_0^t x(t)\, dt \quad \text{or}$$

   $$\tau_i \frac{dy(t)}{dt} + x(t) \tag{2.19}$$

4. *Impure (Imperfect) differentiator*
   With:

   $$y(t) = \tau_d \frac{dx(t)}{dt} + x(t) \tag{2.20}$$

5. *Impure (imperfect) integrator or a first order system*
   With:

   $$\tau_i \frac{dy(t)}{dt} + y(t) = x(t) \tag{2.21}$$

6. *Second order system*
   With:

   $$\frac{1}{\omega_0^2} \frac{d^2 y(t)}{dt^2} + 2\frac{\beta}{\omega_0} \frac{dy(t)}{dt} + y(t) = x(t) \tag{2.22}$$

7. *Time delay*
With:

$$y(t) = x(t - T_d) \tag{2.23}$$

As we shall discuss the second order system repeatedly in the pole/zero theory, we shall elaborate on this basic system. In the system differential equation we find parameters ω_0 and β. These are defined as:

ω_0: The undamped frequency in rad/s of a second order system in which $\beta = 0$.
β: The damping-ratio of a second order system ($\beta \geqq 0$).

If a second order system with damping ratio equal to zero is briefly disturbed at the input (or elsewhere), the output will be sinusoidal with a constant amplitude, in which the damped natural frequency equals ω_0. If $\beta \neq 0$, then the amplitude of the sinusoidal output is reduced by an enveloping e-function. The damped natural frequency in that case equals:

$$\omega_g = \omega_0 \sqrt{1 - \beta^2} \tag{2.24}$$

For $\beta \geqq 1$ there appears no sinusoidal component in the output. Systems with $\beta < 1$ are called under-critically damped, with $\beta = 1$ critically damped, and with $\beta > 1$ over-critically damped.

Example 2.6
The general differential equation of a second order system is:

$$\frac{1}{\omega_0^2} \frac{d^2y(t)}{dt^2} + 2 \frac{\beta}{\omega_0} \frac{dy(t)}{dt} + y(y) = K\,x(t) \tag{2.25}$$

The static gain, or dc gain when an electric system is involved, is K. For $\beta = 0$:

$$\frac{d^2y(t)}{dt^2} + \omega_0^2 y(t) = K \cdot \omega_0^2 x(t) \tag{2.26}$$

Suppose $x(t) = 1(t)$; this is a unit step just after $t = 0$. The characteristic equation is

$$p^2 + \omega_0^2 = 0 \tag{2.27}$$

There are two roots, namely $p_{1,2} = \pm j\omega_0$, with which we obtain the complementary solution of the reduced differential equation.

$$y(t) = C_1 e^{j\omega_0 t} + C_2 e^{-j\omega_0 t}, \text{ with } C_1 = C_2 = C \qquad (2.28)$$
$$= 2C \cos \omega_0 t \qquad (2.29)$$

The particular solution is a constant, so that the output consists of the superposition of a cosinudoisal signal and a constant signal. The frequency of the cosinusoidal signal is ω_0, and the amplitude of this signal is constant.

For $0 < \beta < 1$ the characteristic equation is:

$$p^2 + 2\beta\omega_0 p + \omega_0^2 = 0 \qquad (2.30)$$

Also:

$$p^2 + 2\beta\omega_0 p + \beta^2\omega_0^2 + \omega_0^2 - \beta^2\omega_0^2 = 0$$

$$(p + \beta\omega_0)^2 + \omega_0^2(1 - \beta^2) = 0$$

$$(p + \beta\omega_0 + j\omega_0\sqrt{1-\beta^2})(p + \beta\omega_0 - j\omega_0\sqrt{1-\beta^2}) = 0 \qquad (2.31)$$

The roots of the characteristic equation are:

$$p_{1,2} = -\beta\omega_0 \pm j\omega_0\sqrt{1-\beta^2} \qquad (2.32)$$

The transient in the response becomes:

$$y_1(t) = C_1 e^{-\beta\omega_0 t} \cos(\omega_0\sqrt{1-\beta^2} \cdot t - \varphi) \qquad (2.33)$$

The amplitude of the cosinusoidal signal decreases with a factor: $e^{-\beta\omega_0 t}$, while the damped natural frequency is:

$$\omega_g = \omega_0\sqrt{1-\beta^2} \qquad (2.34)$$

For $\beta \geq 1$ the second order system is split into two identical ($\beta = 1$), or two different ($\beta > 1$) first order systems. In these cases the step response consists of two e-functions. In these cases there is no oscillating transient. The step response for some values of damping ratio β is given in Fig. 2.5.

2.3 SYSTEM DESCRIPTION IN THE ω-DOMAIN

In the frequency domain (ω-domain) the system is described by the complex transfer function. This method may be used only when we consider sinusoidal signals.

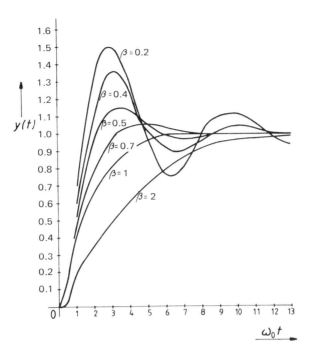

Fig. 2.5 — Step responses of a second order system.

Example 2.7
The system in Fig. 2.6 is similar to the system in Fig. 2.1, but with sinusoidal input and output signals.

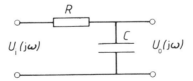

Fig. 2.6 — Electrical first order system.

Considering this system to be a complex voltage divider we find:

$$H(j\omega) = \frac{U_o(j\omega)}{U_i(j\omega)} = \frac{\frac{1}{j\omega C}}{R + \frac{1}{j\omega C}} = \frac{1}{1 + j\omega RC} \qquad (2.35)$$

We can also write (2.35) in the form:

$$RCj\omega U_o(j\omega) + U_o(j\omega) = U_i(j\omega) \tag{2.36}$$

There are apparent similarities between (2.36) and (2.5). By substituting

$$\left.\begin{array}{c} \dfrac{d}{dt} \Leftrightarrow j\omega \\[2ex] f(t) \Leftrightarrow F(j\omega) \end{array}\right\} \tag{2.37}$$

the expressions merge. It can be derived and proved that by the sibstitutions of (2.37) it is possible to shift from a description in the t-domain to a description in the ω-domain, and vice versa. The basic systems that were described in the time domain by the differential equations (2.17) to (2.22) can be described by the transfer functions (2.38) to (2.43) in the ω-domain. The delay, after transforming the time shift T_d into a phase lag, ωT_d, is described by (2.44).

Basic systems:
1. *Constant factor*

$$H(j\omega) = K \tag{2.31}$$

2. *Pure differentiator*

$$H(j\omega) = j\omega\tau_d \tag{2.39}$$

3. *Pure integrator*

$$H(j\omega) = \frac{1}{j\omega\tau_i} \tag{2.40}$$

4. *Imperfect differentiator*

$$H(j\omega) = 1 + j\omega\tau_d \tag{2.41}$$

5. *Imperfect integrator or first order system*

$$H(j\omega) = \frac{1}{1 + j\omega\tau_i} \tag{2.42}$$

6. *Second order system*

$$H(j\omega) = \frac{1}{\left(\dfrac{j\omega}{\omega_0}\right)^2 + 2\beta \dfrac{j\omega}{\omega_0} + 1} \tag{2.43}$$

7. *Time delay*

$$H(j\omega) = e^{-j\omega T_d} \tag{2.44}$$

For the second order system, the frequency characteristics are given in Fig. 2.7, with the damping ratio β as a variable parameter.

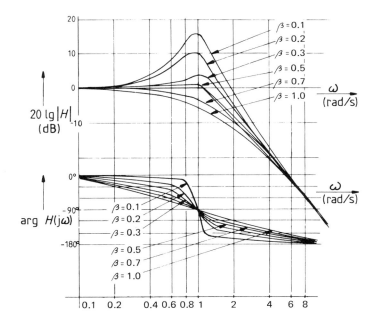

Fig. 2.7 — Frequency characteristics of a second order system.

Note that:
- For $\beta = 0$ the modules of the transfer ratio is infinite at $\omega = \omega_0$; for values of β rising from 0 the maximum of this modulus steadily decreases until for $\beta \geqq 0.7$ the maximum of the modulus is one (0 dB.).
- The position of the occurring maxima shifts to the left for an increasing value of β from $\omega = \omega_0$.
- For $0.7 \leqq \beta \leqq 1$ there is no overshoot in the frequency characteristics, but in the step response overshoot does occur!

Example 2.8
We have a second order system with transfer function:

$$H(j\omega) = \frac{205}{(j\omega)^2 + 8j\omega + 41} \qquad (2.45)$$

Determine the characteristic properties of this second order system.

Solution:
First we put $H(j\omega)$ as in the expression (2.43):

$$H(j\omega) = \frac{5}{\left(\dfrac{j\omega}{\sqrt{41}}\right)^2 + \dfrac{8}{41} j\omega + 1} \qquad (2.46)$$

The dc gain appears to be 5; $\omega_0 = \sqrt{41}$ rad/s; $\beta = \dfrac{4}{41}\sqrt{41} \approx 0.62$; $\omega_g \approx 0.78\omega_0$ rad/s.

For control actions we often apply a tamed differential action and an integral action (I action). The transfer function of a tamed D-action, usually called lead compensation, is described as:

$$H_d(j\omega) = \frac{1 + j\omega\tau_d}{1 + j\omega\dfrac{\tau_d}{a}} \qquad (2.47)$$

for which in practice we usually choose for a the values 6, 10, and 20; such a D-action is tamed in order to reduce the amplification for higher frequencies (e.g. noise). The frequency characteristics are shown in Fig. 2.8; especially important is the phase lead of such a D-action. For maximum phase lead:

$$\varphi_{max} = \arcsin \frac{a-1}{a+1} \qquad (2.48)$$

As an integral control action we usually apply a network with the transfer function:

$$H_i(j\omega) = 1 + \frac{1}{j\omega\tau_i} = \frac{1 + j\omega\tau_i}{j\omega\tau_i} \qquad (2.49)$$

The transfer function of (2.49) has, in respect of a pure I-action, the advantage that no extra phase lag is introduced for higher frequencies (possible instability). The frequency characteristics are shown in Fig. 2.9.

Sec. 2.4] **System description in the s-domain** 25

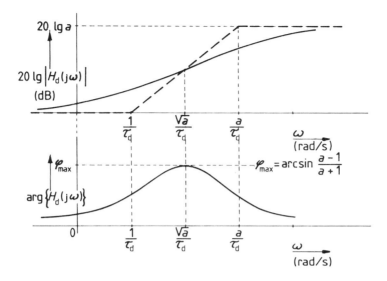

Fig. 2.8 — The Bode diagram of a lead compensation.

Figs 2.10 and 2.11 show possible realizations of a lead compensation and of an I-action by means of operational amplifiers. $H(j\omega)$.

2.4 SYSTEM DESCRIPTION IN THE s-DOMAIN

With the Laplace transform we can determine the transfer function $H(s)$ of a system from the system differential equation. For some Laplace transforms that are used widely in control engineering, and for some calculation rules, see Table 2.1.

The definition formula for the Laplace transform is

$$F(s) = \int_0^\infty f(t)\, e^{-st}\, dt \tag{2.50}$$

Example 2.9

For a system we have this differential equation:

$$2\frac{d^2y(t)}{dt^2} + 12\frac{dy(t)}{dt} + 68\, y(t) = x(t) \tag{2.51}$$

For the input we have:

$$x(t) = 5 \cdot 1(t) \tag{2.52}$$

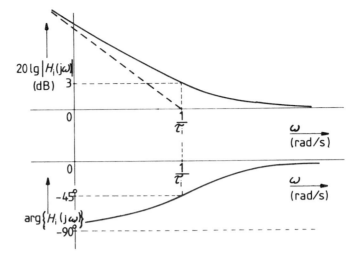

Fig. 2.9 — The Bode diagram of an I-action.

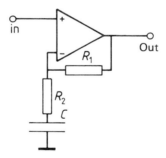

Fig. 2.10 — Realization of a lead network.

The value of $y(t)$ is obtained in the following way.
From (2.51) we have, after Laplace-transform:

$$(2s^2 + 12s + 68)Y(s) = X(s) \ . \tag{2.53}$$

Table 2.1 — Some Laplace transforms and rules: All initial conditions are assumed to be zero

	$f(t)$	$F(s)$
1.	Unit impulses (Dirac pulse) $\delta(t)$	1
2.	Unit Step $1(t)$	$\dfrac{1}{s}$
3.	t	$\dfrac{1}{s^2}$
4.	e^{-at}	$\dfrac{1}{s+a}$
5.	te^{-at}	$\dfrac{1}{(s+a)^2}$
6.	$\sin \omega t$	$\dfrac{\omega}{s^2+\omega^2}$
7.	$\cos \omega t$	$\dfrac{s}{s^2+\omega^2}$
8.	t^n $(n = 1, 2, 3, \ldots)$	$\dfrac{n!}{s^{n+1}}$
9.	$e^{-at}\sin \omega t$	$\dfrac{\omega}{(s+a)^2+\omega^2}$
10.	$e^{-at}\cos \omega t$	$\dfrac{s+a}{(s+a)^2+\omega^2}$

Properties

11.	$\mathcal{L}\{Af(t)\} = AF(s)$	linearity rule
12.	$\mathcal{L}\{f_1(t) \pm f_2(t)\} = F_1(s) \pm F_2(s)$	superposition rule
13.	$\mathcal{L}\left(\dfrac{d^n}{dt^n}f(t)\right) = s^n F(s)$	if $f(0) = f'(0) = \ldots = 0$
14.	$\mathcal{L}\{e^{-at}f(t)\} = F(s+a)$	damping rule
15.	$\mathcal{L}\{f(t-a)\} = e^{-as}F(s)$	shift rule
16.	$\lim\limits_{t \to \infty} f(t) = \lim\limits_{s \to 0} sF(s)$	final value theorem
17.	$\lim\limits_{t \to 0} f(t) = \lim\limits_{s \to \infty} sF(s)$	initial value theorem

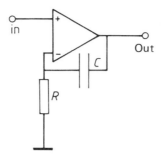

Fig. 2.11 — Realization of an I-action.

Also:

$$X(s) = \frac{5}{s} \quad (2.54)$$

For $Y(s)$ we now have:

$$Y(s) = \frac{5/2}{s(s^2 + 6s + 34)} \quad (2.55)$$

Partial fraction in two reversable transform parts gives:

$$Y(s) = \frac{5}{68}\left\{\frac{1}{s} - \frac{s+6}{s^2+6s+34}\right\} \quad (2.56)$$

or:

$$Y(s) = \frac{5}{68}\left\{\frac{1}{s} - \frac{s+3}{(s+3)^2+5^2} - \frac{3}{(s+3)^2+5^2}\right\} \quad (2.57)$$

Application of the damping rule and the Laplace transform table now yields:

$$y(t) = \frac{5}{68}\left\{1 - e^{-3t}\left(\cos 5t + \frac{3}{5}\sin 5t\right)\right\}1(t) \quad (2.58)$$

For a further discussion of system description in the s-domain please see Chapter 3.

2.5 SYSTEM DESCRIPTION IN THE z-DOMAIN

In principle the behaviour of discrete systems can be described by difference equations, as continous systems are described by differential equations. Yet it has

2.6 PROBLEMS

1. Determine the differential equations of the systems in Fig. 2.12 with input $x(t)$ and output $y(t)$.

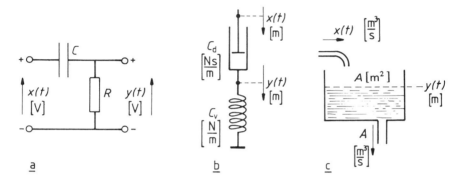

Fig. 2.12 — Some simple processes.

2. A system with input signal $u_i(t)$ and output signal $u_0(t)$ is described by the differential equation:

$$\frac{d^3u_0(t)}{dt^3} + 4\frac{d^2u_0(t)}{dt^2} + 4\frac{du_0(t)}{dt} = 4\frac{du_i(t)}{dt} + 12u_i(t) \qquad (2.59)$$

The input is a step with an amplitude of 5. Determine $u_0(t)$.

3. A system is given by the differential equation:

$$\frac{d^2y(t)}{dt^2} + 8\frac{dy(t)}{dt} + 25y(t) = 4\,x(t) \qquad (2.60)$$

The input is a step of amplitude 25.
Determine $y(t)$ and write the sinusoidal and cosinusoidal terms as one sinusoidal term with a phase shift. Then draw $y(t)$.

4. We have a system with the transform function:

$$H(j\omega) = \frac{3{,}2}{\tfrac{1}{2}(j\omega)^2 + 1{,}6j\omega + 8}$$

Calculate the dc gain, the undamped frequency, the damped frequency, and the damping ratio. Also draw the Bode diagram of this system.

5. Draw the unit step response of the system with the transfer function:

$$H(j\omega) = \frac{4e^{-j\omega^2}}{1+j\omega 2} \qquad (2.61)$$

6. Draw the Bode diagram of the controller with PD-action according to:

$$H_R(j\omega) = 10 \cdot \frac{1+j\omega 0.2}{1+j\omega 0.02}$$

Also find the maximum phase-lead and the frequency at which it occurs.

7. Draw the Bode diagram of the controller with PI action according to:

$$H_R(j\omega) = 2 \cdot \frac{1+j\omega 2}{j\omega 2}$$

8. Find, by means of Laplace transform, the output of the following systems with the indicated input signals $x(t)$:

a $\quad H(s) = \dfrac{4}{s}$; $\qquad x(t) = 2\,1(t)$.

b $\quad H(s) = \dfrac{2s+3}{s^3+2s^2+2s}$; $\qquad x(t) = \delta(t)$.

c $\quad H(s) = \dfrac{4}{s^2+3s+2}$; $\qquad x(t) = 2t\,1(t)$.

d $\quad H(s) = \dfrac{10}{s^3+3s^2+3s+1}$; $\qquad x(t) = \delta(t)$

e $\quad H(s) = \dfrac{5}{s+2}$; $\qquad x(t) = 5\sin 3t\,1(t)$

f $\quad H(s) = \dfrac{2s+7}{s^2+6s+9}$; $\qquad x(t) = 1(t)$

9. We have the time function $x(t)$ as in Fig. 2.13. Determine $X(s)$ both with the definition formula (2.5) and with a graphic composition of three simple functions (step, ramp etc.).

10. We have the transfer function $H(s) = \dfrac{3s+5}{s^3+4s^2+5s+2}$.

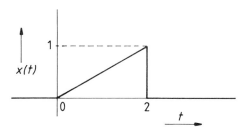

Fig. 2.13 — Ramp signal $x(t)$.

a. Determine, by initial and final value theorems, $c(0)$ and $c(\infty)$, if $c(t)$ represents the impulse response.
b. Determine the impulse response $c(t)$.

11. A thermometer needs 1 minute and 10 seconds to show 97% of a response to a step-like temperature change. We consider the thermometer as a first order system.

   a. Determine the time constant of the thermometer. The thermometer is placed in a bath in which the temperature is rising by a linear function of 12°C/min.
   b. Determine the response of the thermometer.
   c. Determine the error shown after 'a long time'.

3
System description in the s-domain

3.1 INTRODUCTION

If a system is given by the differential equation describing the relation between the input signal and the output signal, it is possible to determine the output signal from a given input signal. Usually the Laplace transform will be used to solve the differential equation. The procedure to determine the output signal is given in Fig. 3.1

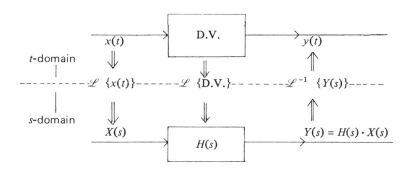

Fig. 3.1 — Procedure of the application of the Laplace transform.

From the differential equation the transfer function $H(s)$ is determined by Laplace transform. The Laplace transformed form of the output signal can then be determined as the product of the transfer function and the Laplace transformed input signal. Through inverse Laplace transform we can determine the output signal $y(t)$ from $y(s)$.

As the differential equation of the system contains all the relevant information for a judgement of the control behaviour of the system, the transfer function will

Sec. 3.2] The poles and zeros presentation

contain this information also. It is also possible to make a useful presentation of $H(s)$ in which the properties of the system are shown. We shall discuss how this presentation develops and the relations with the responses in the t-domain and the above mentioned transfer function $H(j\omega)$.

3.2 THE POLES AND ZEROS PRESENTATION

From the differential equation we determine the transfer function $H(s)$ by Laplace transform. The common form of a differential equation describing a linear time independent system is:

$$a_n \frac{d^n y(t)}{dt^n} + a_{n-1} \frac{d^{n-1} y(t)}{dt^{n-1}} + \ldots + a_1 \frac{dy(t)}{dt} + a_0 y(t) =$$
$$b_m \frac{d^m x(t)}{dt^m} + b_{m-1} \frac{d^{m-1} x(t)}{dt^{m,-1}} + \ldots + b_1 \frac{dx(t)}{dt} + b_0 x(t) \ . \tag{3.1}$$

In this differential equation $x(t)$ is the input signal and $y(t)$ the output signal. Application of the Laplace transform gives, under the condition that all initial conditions are zero, the following equation in s:

$$a_n s^n Y(s) + a_{n-1} s^{n-1} Y(s) + \ldots + a_1 sY(s) + a_0 Y(s) =$$
$$b_m s^m X(s) + b_{m-1} s^{m-1} X(s) + \ldots + b_1 sX(s) + b_0 X(s) \tag{3.2}$$

For the transfer function $H(s)$ we can write:

$$H(s) = \frac{Y(s)}{X(s)} = \frac{b_m s^m + b_{m-1} s^{m-1} + \ldots + b_1 s + b_0}{a_n s^n + a_{n-1} s^{n-1} + \ldots + a_1 s + a_0} \tag{3.3}$$

For a physically realizable system $n \geq m$. It is easy to see that if $n < m$, then the initial condition of the response $y(t)$ to a unit step $x(t) = 1.1(t)$ must be:

$$y(0) = \lim_{s \to \infty} sY(s) = \lim_{s \to \infty} s \cdot \frac{1}{s} \cdot H(s) =$$
$$= \lim_{s \to \infty} \frac{b_m s^m + b_{m-1} s^{m-1} + \ldots + b_1 s + b_0}{a_n s^n + a_{n-1} s^{n-1} + \ldots + a_1 s + a_0}$$
$$= \infty, \text{ if } n < m.$$

A finite input signal would cause infinity immediately after $t = 0$: no physical system shows such properties. Also, in another way, we can see that for a physical system $n \geq m$. From (3.3) follows by substitution $s = j\omega$ the transfer function $H(j\omega)$:

$$H(j\omega) = \frac{Y(j\omega)}{X(j\omega)} = \frac{b_m(j\omega)^m + b_{m-1}(j\omega)^{m-1} + \ldots + b_1 j\omega + b_0}{a_n(j\omega)^n + a_{n-1}(j\omega)^{n-1} + \ldots + a_1 j\omega + a_0} \qquad (3.4)$$

For $m > n$, $|H(j\omega)|$ approaches infinity if $\omega \to \infty$. This would mean that such a system would give an ever higher amplification of input signals of a higher frequency; no physical system shows this property.

In (3.3) the polynomial found in s can always be written as:

$$H(s) = \frac{b_m}{a_n} \cdot \frac{s^m + \frac{b_{m-1}}{b_m} s^{m-1} + \frac{b_{m-2}}{b_m} s^{m-2} + \ldots + \frac{b_1}{b_m}}{s + \frac{b_0}{b_m} s^n + \frac{a_{n-1}}{a_n} s^{n-1} + \frac{a_{n-2}}{a_{n2}} s^n + \ldots + \frac{a_1}{a_n} s + \frac{a_0}{a_n}} \qquad (3.5)$$

$$= \frac{b_m}{a_n} \cdot \frac{(s-z_1)(s-z_2)\ldots(s-z_m)}{(s-p_1)(s-p_2)\ldots(s-p_n)}$$

$$= \frac{b_m}{a_n} \prod_{i=1}^{m} (s - z_i) \prod_{i=1}^{n} (s - p_i) \qquad (3.6)$$

The factor b_m/a_n is called the pole-zero gain K_{pn}. Notice that this factor develops from giving a value of 1 to the factors in $H(s)$ that are linked to the highest degree of s. The value of K_{pn} is certainly not equal to the dc gain K of the system, for this gain is $K = b_0/a_0$.

The values of s for $H(s) = 0$ we find from:

$$\prod_{i=1}^{m} (s - z_i) = 0 \qquad (3.7)$$

These values of s are called the zeros of the transfer function $H(s)$. For these zeros:

$$s = z_i, \text{ with } i = 1, 2, 3, \ldots, m. \qquad (3.8)$$

There are as many zeros as the degree of the numerator polynomial of $H(s)$, that is, m.

The values of s that are in $H(s) = \infty$, we find from:

$$\prod_{i=1}^{n} (s - p_i) = 0 \qquad (3.9)$$

These values are called the poles of $H(s)$. For these poles:

Sec. 3.2] The poles and zeros presentation

$$s = p_i, \text{ with } i = 1,2,3,\ldots, n. \tag{3.10}$$

There are as many poles as the denominator polynomial of $H(s)$, that is, n. The poles and zeros can be real or complex; complex poles and/or zeros always occur as additional complex pairs, for the transfer function itself is always real.

As the transfer function describes the system as a whole, the conditions of the pole/zero gain and of the zeros and the poles will contain these data values, for we always find:

$$H(s) = K_{\text{pn}} \prod_{i=1}^{m} (s - z_i) \prod_{i=1}^{n} (s - p_i) \Longleftrightarrow \text{pn-presentation} \tag{3.11}$$

The poles and zeros can be described in the complex s-plane. Therefore we put:

$$s = \lambda + j\omega \tag{3.12}$$

So the real part of s is λ, and the imaginary part ω. This choice is made on purpose, because from $\lambda = 0$ develops the above determined relation $s = j\omega$ in which ω is the (real) frequency, expressed in radian per second. In this connection s is also called the complex frequency. As we shall see later, λ also has a certain physical meaning.

The position of a pole is indicated by a cross ($\times$) and of a zero by a circle ($\bigcirc$); if poles or zeros coincide, this is indicated by giving the numbers between brackets. In the literature we also find different methods of notation, such as ✬ for two identical poles and ⊚ for two identical zeros. In the upper left hand corner of the pole/zero plot the value of K_{pn} must be given.

Example 3.1
A system is described by the following differential equation:

$$2 \frac{d^2 y(t)}{dt^2} + 14 \frac{dy(t)}{dt} + 20\, y(t) = 10 \frac{dx(t)}{dt} + 30\, x(t)$$

$x(t)$ is the input signal and $y(t)$ is the output signal.
We are asked to determine the pole/zero plot of this system.

Solution:
The transfer function is:

$$H(s) = \frac{10\, s + 30}{2\, s^2 + 14\, s + 20} = 5 \cdot \frac{s + 3}{(s + 2)(s + 5)}$$

The pole-zero gain is 5, there is one zero $s_1 = -3$, and there are two poles, namely $s_2 = -2$, and $s_3 = -5$; the pole/zero plot is shown in Fig. 3.2.

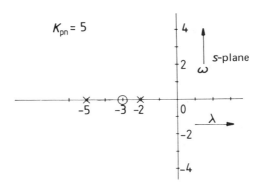

Fig 3.2 — The pole/zero plot of $H(S)$.

Example 3.2
A system is given by the following transfer function:

$$H(s) = \frac{2s+4}{s^3 + 2s^2 + 3s}$$

Draw the pole/zero plot of $H(s)$

Solution:

Here: $H(s) = 2\dfrac{s+2}{s(s^2+2s+3)} = 2 \cdot \dfrac{s+2}{s\{(s+1)^2 + 2\}}$

$= 2\dfrac{s+2}{s(s+1+j\sqrt{2})(s+1-j\sqrt{2})}$

The pole/zero plot is shown in Fig. 3.3.

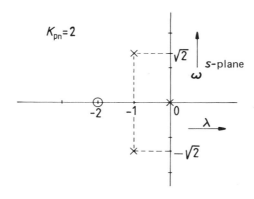

Fig. 3.3 Pole/zero plot of $H(s)$.

Sec. 3.3]   **The pole/zero plot and responses in the *t*-domain**   37

Example 3.3
The transfer function of a process is given as follows:

$$H(s) = \frac{4s^2 + 16s + 32}{3s^4 + 12s^2}$$

Draw the pole/zero plot of this process.

Solution
$H(s)$ can be written as:

$$H(s) = \frac{4}{3} \frac{(s+2+j2)(s+2-j2)}{s^2(s+j2)(s-j2)}$$

The pole zero/plot is as shown in Fig. 3.4.

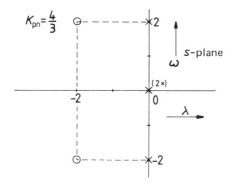

Fig. 3.4 — Pole/zero plot of $H(s)$.

Example 3.4
Fig. 3.5 gives the pole/zero plots of some simple systems. Please check these.
In the next paragraph it will become clear that the position of the poles and zeros of the system are determining the behaviour of the system.

3.3 THE RELATION BETWEEN THE POLE/ZERO PLOT AND RESPONSES IN THE *t*-DOMAIN

As indicated in Fig. 3.1, a response of a system can be determined by applying the inverse Laplace transform to $Y(s)$. To achieve this, $Y(s)$ must be so adapted that $Y(s)$ is split into a number of factors, and that each of them can be transformed inversely

38   System description in the *s*-domain   [Ch. 3

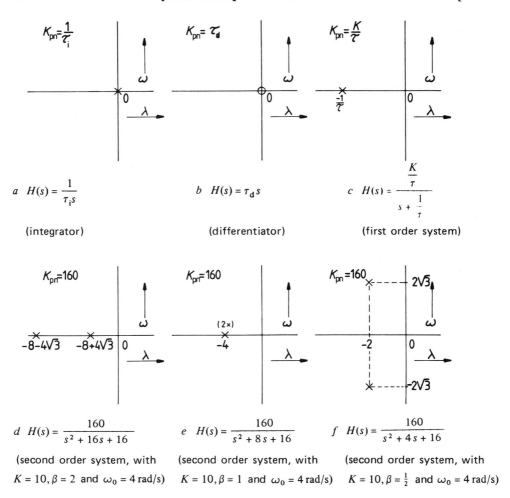

Fig. 3.5 — Pole/zero plots of some simple systems.

by the Laplace transform table. There is the obvious question whether it is possible to determine a certain time response from pole/zero plot of a system. This appears to be the case, and from a quality point of view it is rather easy. Quantitatively such a determination is considerably more difficult, as we see later.

For the determination of a time response we must take into account both the properties of the system and the nature of the input signal. Therefore we draw the pole/zero plot of a system first and then add the parameters of the input signal. Here $Y(s) = H(s) \cdot X(s)$, and $H(s)$ and $X(s)$ can be split into a constant factor and some poles and/or zeros.

Example 3.5
A system is described by the following differential equation:

Sec. 3.3] The pole/zero plot and responses in the t-domain

$$2\frac{d^3 y(t)}{dt^3} + 14\frac{d^2y(t)}{dt^2} + 20\frac{dy(t)}{dt} = 4\frac{dx(t)}{dt} + 12x(t)$$

The input signal is: $x(t) = 5.1(t)$.
Now determine the pole/zero plot of $Y(s)$

Solution:

Here: $H(s) = 2 \cdot \dfrac{s+3}{s(s+2)(s+5)}$

$X(s) = \dfrac{5}{s}$

$Y(s) = 10 \cdot \dfrac{s+3}{s^2(s+2)(s+5)}$

and the pole/zero plot is given in Fig. 3.6.

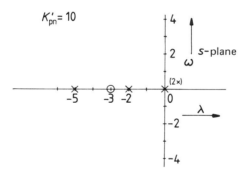

Fig. 3.6 — Pole/zero plot of $Y(s)$.

Example 3.6
A system is described by the transfer function:

$$H(s) = -4\frac{s}{s^2 + 8s + 17}$$

The input signal is $x(t) = 4 e^{-t} \cos 2t \cdot 1(t)$.
Determine the pole/zero plot of $Y(s)$.

Solution:

$$X(s) = 4\frac{s+1}{(s+1)^2 + 4}$$

$$Y(s) = -16\frac{s(s+1)}{(s+4+j)(s+4-j)(s+1+j2)(s+1-j2)}$$

The pole/zero plot is given in Fig. 3.7.

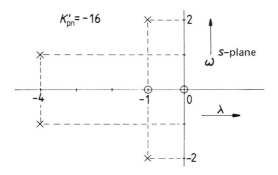

Fig. 3.7 — Pole/zero plot of $Y(s)$.

For the Laplace transformed form of the output signal generally:

$$Y(s) = H(s) \cdot X(s) = K'_{pn} \cdot \frac{T'(s)}{N'(s)} \qquad (3.13)$$

in which the highest degree of s in the numerator (u) is less than the highest degree of s in the denominator (v); under this condition $Y(s)$ can always be written as:

$$Y(s) = K'_{pn}\left\{\frac{A_1}{s-p_1} + \frac{A_2}{s-p_2} + \ldots + \frac{A_v}{s-p_v}\right\} \qquad (3.14)$$

Also:

$$y(t) = K'_{pn}\left\{A_1 e^{p_1 t} + A_2 e^{p_2 t} + \ldots + A_v e^{p_v t}\right\} \qquad (3.15)$$

The signal $y(t)$ is the summation of a number of e-functions in which the exponents

The pole/zero plot and responses in the *t*-domain

are determined by the position of the poles. We shall see that in the factors A_1 to A_ν the influence of the position on the (possible) zeros is important.

For a complete determination of $y(t)$ we still have to find these factors A_1 to A_ν. From (3.13) and (3.14) follows:

$$\frac{\prod_{i=1}^{u}(s-z_i)}{\prod_{i=1}^{\nu}(s-p_i)} = \frac{A_1}{s-p_1} + \frac{A_2}{s-p_2} + \ldots + \frac{A_\nu}{s-p_\nu} \qquad (3.16)$$

Multiplication of (3.16) by $(s-p_1)$ gives:

$$\frac{\prod_{i=1}^{u}(s-z_i)}{\prod_{i=2}^{\nu}(s-p_i)} = A_1 + (s-p_1)\left\{\frac{A_2}{s-p_2} + \ldots + \frac{A_\nu}{s-p_\nu}\right\}$$

If we now give s a value of p_1, we find:

$$A_1 = \frac{\prod_{i=1}^{u}(p_1-z_i)}{\prod_{i=2}^{\nu}(p_1-p_i)} \qquad (3.17)$$

To determine any factor A_j we can use the same equation (3.16), resulting in:

$$A_j = \frac{\prod_{i=1}^{u}(p_j-z_i)}{\prod_{\substack{i=1\\i\neq j}}^{\nu}(p_j-p_i)} \qquad (3.18)$$

In Fig 3.8 the example expresses the form of $(p_j - z_i)$, considering a zero z_i and a pole p_j; the expression $p_j - z_i$ appears to be the vector from the zero z_i to the pole p_j.

Thus formula (3.18) has this meaning:

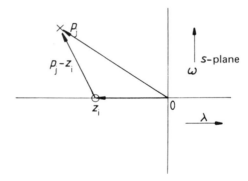

Fig. 3.8 — Construction of the vector $(p_j - z_i)$.

The factor A_j according to the e-function in which the pole p_j is expressed in the time response, can be determined as being the quotient of the product of the vectors from all zeros to the pole p_j and the product of the vectors from all other poles to pole p_j.

Example 3.7
A system is described by the differential equation:

$$2 \frac{dy(t)}{dt} + 8y(t) = 10x(t)$$

The input signal is a unit step. Determine the response $y(t)$ from the pole/zero plot.

Solution:
The transfer function becomes:

$$H(s) = \frac{5}{(s+4)}.$$

The input signal becomes:

$$X(s) = \frac{1}{s}.$$

For $Y(s)$:

$$Y(s) = \frac{5}{s(s+4)}.$$

The pole/zero plot becomes: (Fig. 3.9a)

[Sec. 3.3] **The pole/zero plot and responses in the t-domain**

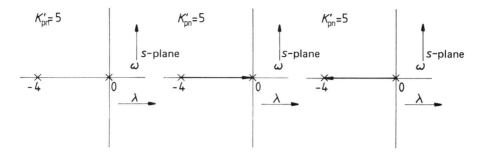

Fig. 3.9a — Pole zero of Y(s)   Fig. 3.9b — Determination of A_1.   Fig. 3.9c — Determination of A_2.

The response becomes:

$$y(t) = 5\{A_1 e^{0t} + A_2 e^{-4t}\},$$

with

$$A_1 = \frac{1}{+4},$$

see Fig. 3.9b and

$$A_2 = \frac{1}{-4},$$

see Fig. 3.9c.
We now obtain:

$$y(t) = \frac{5}{4}\{1 - e^{-4t}\}\, 1(t)$$

Example 3.8
A system has the following transfer function:

$$H(s) = 3\frac{s+5}{(s+2)(s+9)}$$

Determine the impulse response.

Solution:
For $x(t) = \delta(t)$, $X(s) = 1$, and so $Y(s) = H(s)$.
The pole/zero plot is shown in Fig. 3.10.

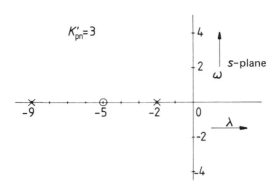

Fig. 3.10 — Pole/zero plot of $Y(s)$.

For $Y(t)$

$$y(t) = 3\{A_1 e^{-2t} + A_2 e^{-9t}\}, \text{ with}$$

$$A_1 = \frac{+3}{+7} \text{ and } A_2 = \frac{-4}{-7} = \frac{4}{7}$$

or $\quad y(t) = \dfrac{3}{7}\{3e^{-2t} + 4e^{-9t}\}\, 1(t)$

Example 3.9
For a second order system:

$$H(s) = \frac{20}{s^2 + 8s + 32}$$

At $t = 0$ we put a step disturbance of 5 on the input. Determine the response.

Solution:
For $Y(s)$:

$$Y(s) = \frac{100}{s(s+4+j4)(s+4-j4)},$$

from which the pole/zero plot is as in Fig. 3.11.
Now we have: $y(t) = 100\{A_1 e^{0t} + A_2 e^{(-4+j4)t} + A_3 e^{(-4-j4)t}\}$

and $\quad A_1 = \dfrac{1}{(4-j4)(4+j4)} = \dfrac{1}{32}$

Sec. 3.3] The pole/zero plot and responses in the *t*-domain

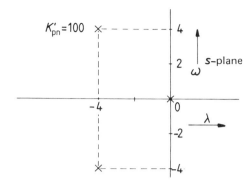

Fig. 3.11 — Pole/zero plot of $Y(s)$.

$$A_2 = \frac{1}{(-4+j4)\cdot j8} = \frac{1}{32} \cdot \frac{1}{-1-j} = -\frac{1}{64} \cdot (1-j)$$

$$A_3 = -\frac{1}{64}(1+j)$$

From the above we have:

$$y(t) = \frac{25}{16}\left\{2 - (1-j)\,e^{(-4+j4)t} - (1+j)\,e^{-(4-j4)t}\right\} =$$

$$= \frac{25}{16}\left[2 - e^{-4t}\left\{(1-j)e^{j4t} + (1+j)e^{-j4t}\right\}\right] =$$

$$= \frac{25}{16}\left[2 - e^{-4t}\left\{(e^{j4t} + e^{-j4t}) - j(e^{j4t} - e^{-j4t})\right\}\right] =$$

$$= \frac{25}{8}\left\{1 - e^{-4t}(\cos 4t + \sin 4t)\right\}$$

Some remarks on the procedure in the examples:

- With coinciding poles a difficulty occurs because the procedure does not lead simply to the desired result. A practical approach is to separate two or more coinciding poles a little, and then to apply the rules. Of course it is also possible not to start from the representation of $Y(s)$ but to apply the partial fraction method followed by inverse Laplace transform.
- If there are no zeros we must substitute a value of 1 in the numerator to determine the factor A in the response; this results from the earlier agreement for K_{pn}, (the factor in s^o becomes 1 in this case) in the expression (3.13).

- If there are e-functions in the response with complex conjugated poles, then also the A factors belonging to them are complex conjugated. In e-functions with real poles also the A factors are real. These considerations are easily understandable when we realize that the response $Y(t)$ is of course real.
- If a response must be described it is often easier to begin with one sinus or cosinus with (possible) phase shift than with a number of sinusoidal and/or cosinusiodal terms. To achieve this we can determine the sum of such terms as follows:

$$y(t) = 10\, e^{-4t} (\cos 5t + 3 \sin 5t)$$

Now put $\tan \varphi = 3$, then we have:

$$y(t) = 10\, e^{-4t} \frac{1}{\cos \varphi} (\cos 5t \cos \varphi + \sin 5t \sin \varphi)$$
$$= 10 \sqrt{10} \cdot e^{-4t} \cos(5t - \varphi)$$

or with $\cot \varphi = 3$, then we obtain:

$$y(t) = 10\, e^{-4t} \frac{1}{\sin \varphi} (\cos 5t \sin \varphi + \sin 5t \cos \varphi)$$
$$= 10 \sqrt{10}\, e^{-4t} \sin(5t + \varphi).$$

As we have seen from the examples, the quantitative determination of the response from the pole/zero plot is rather laborious, especially when complex poles and/or zeros occur in $Y(s)$. Qualitatively, however, we can soon find a response from the pole/zero plot of $Y(s)$.

In Fig. 3.12 a number of pole positionings are indicated, together with the property of the step response. Of course, complex poles continuously occur in complex conjugated poles; in the figures there is only one of them at a time. It is clear that when poles are positioned further away from the imaginary axis, the transient of the step response will be zero sooner. We can also say the damping is larger when the poles are more to the left in the s-plane. Poles on the imaginary axis lead to permanent components in the response. These components are undamped. When there are poles in the right side of the half plane of the s-plane, these poles lead to components in the response that become ever larger. Later on we shall discuss the nature of the above mentioned damping.

If one or more poles are far to the left in the left-hand half plane with respect to other poles of the system, their part in the total response will be of short duration. It is important, too, that the amplitude of this contribution is small, because the A factors according to (3.18) are small.

Sec. 3.4]   **Pole/zero representation and the transfer function**   47

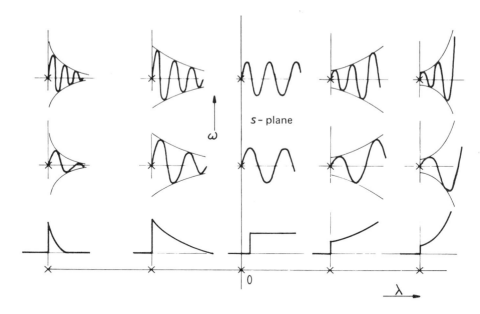

Fig. 3.12 — Some responses in connection with the pole positions.

3.4 THE RELATIONSHIP BETWEEN THE POLE/ZERO REPRESENTATION AND THE TRANSFER FUNCTION

In the ω-domain, the most frequently used representations are formed by the polar figure and the Bode diagram. An important role is played by $|H(j\omega)|$ and $\arg\{H(j\omega)\}$. Here $H(j\omega)$ is the complex transfer function of the system. Notice that at the transfer from the s-domain to the ω-domain only the information concerning the steady state of, for example the output is preserved. It is a fact the $H(j\omega)$ only renders information about amplitude and phase shift of responses to sinusoidal input signals.

We have already discussed the fact that the relation between the s-domain and the ω-domain is given by the substitution $s = j\omega$. The frequencies concerning the polar figure and the Bode diagram are indicated on the positive imaginary axis in the s-domain. It will become clear that $|H(j\omega)|$ and $\arg\{H(j\omega)\}$ can easily be determined from the pole/zero plot of the transfer function of a system.

Here we begin with the general description of $H(s)$ as mentioned in (3.11):

$$H(s) = K_{pn} \frac{\prod_{i=1}^{m}(s - z_i)}{\prod_{i=1}^{n}(s - p_i)}$$

For $s = j\omega$ we obtain:

$$H(j\omega) = K_{pn} \frac{\prod_{i=1}^{m}(j\omega - z_i)}{\prod_{i=1}^{n}(j\omega - p_i)} \qquad (3.19)$$

For the modulus of the transfer function we find:

$$|H(j\omega)| = |K_{pn}| \frac{\prod_{i=1}^{m}|j\omega - z_i|}{\prod_{i=1}^{n}|j\omega - p_i|} \qquad (3.20)$$

For $\omega = \omega_1$, $(j\omega_1 - z)$ and $(j\omega_1 - p_i)$ represent the vectors from the zero z_i to the frequency ω_1, and from the pole p_i to the same frequency respectively.

The modulus of the transfer function for $\omega = \omega_1$ is formed by multiplying the modulus of K_{pn} by the product of the lengths of the vectors from the zeros to ω_1 and dividing this result by the product of the lengths of the vectors from the poles to ω_1.

From (3.19) for the argument of the transfer function it follows that:

$$\arg\{H(j\omega_1)\} = \arg\{K_{pn}\} + \sum_{i=1}^{n} \arg\{j\omega_1 - z_i\} - \sum_{i=1}^{n} \arg\{j\omega_1 - p_i\} \qquad (3.21)$$

The argument of the transfer function for $\omega = \omega_1$ develops from adding the argument of K_{pn} to the sum of the arguments of the vectors from the zeros to ω_1, and to subtract from this result the sum of the arguments of the vectors from the poles to ω_1.

Example 3.10
A system is described by the differential equation:

$$\frac{dy(t)}{dt} + 5\, y(t) = 10\, x(t).$$

Determine $|H(j\omega)|$ and $\arg\{H(j\omega)\}$ for $\omega = 0$ and $\omega = 5$ rad/s, from the pole/zero plot of $H(s) = \frac{Y(s)}{X(s)}$.

Solution

$H(s) = \dfrac{10}{s+5}$, gives the pole/zero plot as in Fig. 3.13.

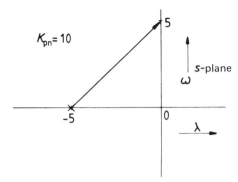

Fig. 3.13 — Pole/zero plot of Example 3.10.

The dc gain of this system is found for $\omega = 0$, so:

$$|H(j0)| = 10 \cdot \frac{1}{5} = 2; \quad \arg\{H(j0)\} = 0$$

Notice that in this simple case we can also derive this conclusion directly from $H(s)$.

For $\omega = 5$ rad/s the vector has been drawn in Fig. 3.13 from the pole in $s = -5$ to $\omega = 5$ rad/s.

Now we find: $\quad |H(j5)| = 10 \cdot \dfrac{1}{5\sqrt{2}} = \sqrt{2}$

and

$$\arg\{H(j5)\} = 0° + 0° - 45° = -45°.$$

For $\omega = \infty$, $|H(j\infty)| = \dfrac{10}{\infty} = 0$ and $\arg\{H(j\infty)\} = 0° + 0° - 90° = -90°$.

The reader will recognize these parameters as characteristic parameters of the first order system, with a dc gain of 2 and a time constant of $\frac{1}{5}$ second.

Example 3.11
A system is given by the transfer function:

$$H(s) = -3 \cdot \frac{s+5}{s(s+9)}$$

Now we must determine $|H(j\omega)|$ and $\arg\{H(j\omega)\}$ for $\omega = 0$ rad/s and $\omega = 13$ rad/s from the pole/zero plot.

Solution:
The pole/zero plot will be as in Fig. 3.14

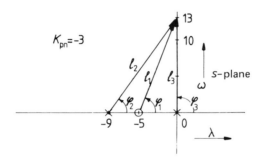

Fig. 3.14 — Pole/zero plot of Example 3.11.

For $\omega = 0$, $|H(j\omega)|$ becomes equal to infinity, because the length of the vector from the pole in the origin to $\omega = 0$ is zero.
The argument for $\omega = 0$ now becomes:

$$\arg\{H(j0)\} = 180° + 0° - 0° - 90° = +90°,$$

because for $\omega \downarrow 0$ the vector from the origin to ω keeps pointing vertically upwards in the *s*-plane.

For $\omega = 13$ rad/s we could determine $|H(j13)|$ by measurement, then we find:

$$|H(j13)| = 3 \cdot \frac{l_1}{l_2 \cdot l_3} \quad \text{and} \quad \arg\{H(j13)\} = 180° + \varphi_1 - \varphi_2 - \varphi_3.$$

These lengths can also be determined by means of Pythagoras' proposition; then we find

$$|H(j13)| = 3 \cdot \frac{\sqrt{194}}{\sqrt{250} \cdot 13} \approx 0.2$$

Sec. 3.4]   Pole/zero representation and the transfer function

The argument follows from:

$$\arg\{H(j13)\} = 180° + \arctan\frac{13}{5} - \arctan\frac{13}{9} - 90° \simeq 104° \ .$$

It goes without saying that, if we measure in the s-plane, the scales of the λ and ω axis must be equal.

Example 3.12
Given the transfer function:

$$H(s) = \frac{100}{s^2 + 6s + 25} = \frac{100}{(s + 3 + j4)(s + 3 - j4)}$$

Now determine the gain and the phase shift for $\omega = 0$ rad/s and $\omega = 4$ rad/s from the pole/zero plot.

Solution:
The pole/zero plot is as given in Fig. 3.15.

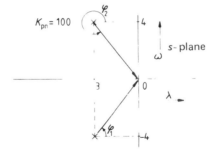

Fig. 3.15 — Pole/zero plot of Example 3.12.

For $\omega = 0$:

$$|H(j0)| = 100 \times \frac{1}{5 \times 5} = 4$$

and $\quad \arg\{H(j0)\} = 0 + 0 - \varphi_1 - \varphi_2 = -360° \equiv 0° \ .$

For $\omega = 4$ rad/s:

$$|H(j4)| = 100 \times \frac{1}{3 \times \sqrt{3^2 + 8^2}} \simeq 3.9 \ .$$

$$\arg\{H(j4)\} = 0 + 0 - 0 - \arctan\frac{8}{3} \simeq -69.4° \ .$$

Also relevant for the determination of the modulus and the argument of $H(j\omega)$ from the pole/zero plot of $H(s)$ is the remark made earlier that the quantitative determination is rather difficult to achieve, but that it is rather easy for the quantitative one. The following conclusions concerning the position of certain poles in the s-plane and the frequency behaviour of the system, the reader may check for himself:

- A zero in the origin gives a constant phase lead of 90°.
- A pole in the origin gives a constant phase lag of 90°.
- Two complex conjugated poles close to the imaginary axis in the left half of the s-plane, may cause an emphasis in the Bode diagram.
- Two complex conjugated poles ($s = \pm j\omega_1$) on the imaginary axis enlarge the gain for $\omega = \omega_1$ infinitely. In that case for $\omega = \omega_1$ the damping is zero and there is a spontaneous sinusoidal output signal with $\omega = \omega_1$ and a constant amplitude. The system is called an oscillator. From a control point of view such a system cannot be used; see also Chapter 4.
- A pole/zero combination on the real axis has a decreasing influence on $|H(j\omega)|$ and $\arg\{H(j\omega)\}$ for an increasing value of ω.

3.5 PROBLEMS

1. Carry out the procedure indicated in Fig. 3.1 to determine $y(t)$ if the system is a first order system with a dc gain of 10 and a time constant of 0.1 second. The input signal is a ramp with a slope of 5.
2. Draw the pole/zero plots of the systems given below:

a $$H(s) = \frac{40s + 200}{4 s^3 + 20 s^2 + 24 s}$$

b $$H(s) = \frac{3 s^2 + 18 s + 27}{2 s^3 + 8 s^2 + 3 s + 12}$$

c $$H(s) = \frac{4 s}{s^3 + 3 s^2 + 3 s + 1}$$

d $$H(s) = 4 \frac{s^2 + s + 1}{s^3 - 1}$$

e $$H(s) = 2 \frac{s^2 - 1}{s^3 + 10 s^2 + 9 s}$$

f $$H(j\omega) = \frac{1 + j\omega\frac{1}{2}}{1 + j\omega\frac{1}{20}} \quad \text{(lead compensation)}$$

g $H(j\omega) = 10 \dfrac{1+j\omega}{1+j\omega 10}$ (lag compensation)

h A second order system with $K = 10$, $\omega_0 = 0.01$ rad/s, and $\beta = \tfrac{1}{2}\sqrt{2}$.

i A third order system with three identical time constants of $\tfrac{1}{3}$ second and a total dc gain of 15.

j $\dfrac{d^3y(t)}{dt^3} + 8\dfrac{d^2y(t)}{dt^2} + 12\dfrac{dy(t)}{dt} = 4\dfrac{dx(t)}{dt}$

$x(t)$ is the input signal and $y(t)$ is the output signal.

3. Obtain, by means of the pole/zero plot of $y(s)$, the response $y(t)$ in the following cases:

a A first order system with $K = 10$ and $\tau = 5$ s; the input is a step with an amplitude of 6.

b A second order system with $K = 0$, $\omega_0 = 2$ rad/s, and $\beta = 0.1$; the input is an impulse.

c A third order system with $K = 9$, $\tau_1 = 1$ s, $\tau_2 = 0.1$ s, and $\tau_3 = 0.01$ s; the input is a unit step.

d A first order system with $K = 4$ and $\tau = \tfrac{1}{5}$ s; the input is $x(t) = 4\,e^{-t}\sin 3t \cdot 1(t)$.

4. Obtain, from the pole/zero plot of the systems below, the gain and the phase shift for the frequencies indicated:

a A first order system with $K = 10$, $\tau = 2$ s; $\omega = 0$, $\tfrac{1}{2}$, and ∞ rad/s.

b A second order system with $K = 2$, $\omega_0 = 1$ rad/s, and $\beta = 1$; $\omega = 0, 10,$ and ∞ rad/s.

c A lag compensation network, $H(j\omega) = a \cdot \dfrac{1+j\omega\tau_i}{1+j\omega a\tau_i}$, with a dc gain of 100 and highest corner frequency of 100 rad/s; $\omega = 0, 10,$ and ∞ rad/s.

d A lead compensation network, $H(j\omega) = \dfrac{1+j\omega\tau_d}{1+j\omega\dfrac{\tau_d}{a}}$, with a high frequency gain of 24 dB and a lowest corner frequency of 20 rad/s; $\omega = 0, 80,$ and ∞ rad/s.

4

System analysis in the *s*-plane

4.1 INTRODUCTION

For some time the Laplace transform was only used as a means to solve differential equations. The application of initial and final theorems already indicates the effort to postpone the inverse Laplace transformation and to draw conclusions from, for example the Laplace transformed response of a system. As we have seen in Chapter 3, the position of a pole or zero in the s-plane determines the influence of such a pole or zero on the system's behaviour. It is obvious to find specific criteria in the *s*-domain, just like in the ω-domain, by which the system can be assessed from a control point of view. In the frequency domain we recognize, for example, bandwidth, dc gain and Nyquist's stability criterion. When designing control systems in the ω-domain we can add, for example, a phase- and/or gain margin. Also in the *s*-plane it appears to be possible to describe and indicate the specific criteria by which the system is determined. From the nature of the Laplace transform it follows directly that the criteria used are closely related to the *t*-domain. This appears to be a great advantage of applying control considerations in the *s*-domain.

4.2 STABILITY CRITERIA IN THE *s*-PLANE

According to expression (3.15) the response of a specific system consists of the superposition of a number of factors in the form:

$$A_i \, e^{p_i t} \tag{4.1}$$

The factor A_i is real — if p_i is real — or complex — if p_i is a complex conjugate and expresses the quantitative influence of the related position of the poles and zeros. The position of the pole is further expressed in the exponent of the relevant e-function and so determines the qualitative behaviour of the system. Generally p_i is complex, so (4.1) can also be written as:

Sec. 4.2]   **Stability criteria in the s-plane**   55

$$A_i \, e^{(\lambda_i + j\omega_i)t} = A_i \, e^{\lambda_i t} \, e^{j\omega_i t}, \text{ with } p_i = \lambda_i + j\omega_i \tag{4.2}$$

The e-functions with the imaginary exponents result in an oscillating response; the e-functions with real exponents give, if they originate from the real part of complex poles in the left half-plane, a certain damping of this oscillating response. If a pole is real, an e-function with real exponents may occur on its own; and then also a decreasing signal arises when the relevant pole is in the left half-plane.

Poles in the right half-plane do not have a damping influence with time, but, on the contrary, they have an increasing influence on the total response. Oscillating components will then increase and the value of components of real poles will also be increasing, as we have already seen in Chapter 3.

For poles on the real axis the damping factor is zero. That is, the oscillating phenomenon in the response has a constant amplitude.

One pole in the origin gives a step contribution to the response, and two poles a ramp.

The stability of a system can be defined as follows:

A system is stable only when the result of a limited amplitude input signal is a likewise limited output signal.

As the reader can see for himself from the above observation about damping, from this definition it follows that for a system to be stable all poles of the system must be to the left of the imaginary axis, with the exception of systems that have one pole in the origin. Such (integrating) systems belong to the stable systems provided that all other poles are to the left of the imaginary axis.

Example 4.1
A system is given by the following differential equation:

$$\frac{d^3y(t)}{dt^3} + 4\frac{d^2y(t)}{dt^2} + 8\frac{dy(t)}{dt} = 4\frac{dx(t)}{dt} - 12\,x(t)$$

Is this system stable or not?

Answer:

$$H(s) = \frac{4s - 12}{s^3 + 4s^2 + 8s} = 4\frac{s - 3}{s(s + 2 + j2)(s + 2 - j2)}$$

The pole/zero plot is shown in Fig. 4.1.
The system is stable because all poles are to the left of the imaginary axis, except the one in the origin.

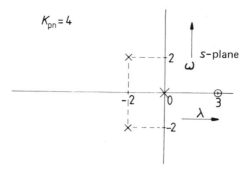

Fig. 4.1 — Pole/zero plot of Example 4.1.

Example 4.2
The transfer function of a system is given by:

$$H(s) = 5\frac{s}{s^2 + 9}$$

Is the system stable or not?

Answer:

Here applies: $H(s) = 5\dfrac{s}{(s + j3)\,(s - j3)}$, with the pole/zero plot of Fig. 4.2.

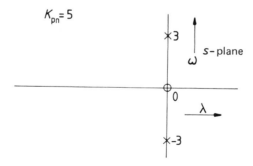

Fig. 4.2 — Pole/zero plot of Example 4.2.

This pole/zero plot shows an unstable system.
 Notice that in assessing the system we must consider the position of the poles only. Poles resulting from the input signal must not be considered. There are also no

Stability criteria in the s-plane

restrictions imposed by the stability criterion, as to the position of the zeros of the system.

For the applied stability criterion it is necessary to know the position of the poles. To achieve this, the denominator polynomial of the transfer function must be split into terms of the form $s - p_i$. It is often difficult to determine the roots of polynomials with a greater degree of s than two. Although, in this case, numerical methods present a solution, there is also an analytical method to test the stability, — Routh's stability criterion. For the application of this criterion we consider the denominator polynomical to be:

$$N(s) = a_n s^n + a_{n-1} s^{n-1} + \ldots + a_1 s + a_0$$

$$= a_n \{s^n + c_{n-1} s^{n-1} + c_{n-2} s^{n-2} + \ldots + c_1 s + c_0\} \qquad (4.3)$$

The requirements for the stability of a system are:

1. All factors (c_{n-1}) up to and including c_0 must be positive.

2. • For third order systems

$$c_1 c_2 - c > 0 \qquad (4.4)$$

• For fourth order systems

$$c_1 c_2 c_3 - c_0 c_3^2 - c_1^2 > 0 \qquad (4.5)$$

• For fifth order systems

$$c_3 c_4 - c_2 > 0 \qquad (4.6)$$

and $(c_3 c_4 - c_2)(c_1 c_2 - c_0 c_3) - (c_1 c_4 - c_0)^2 > 0$

• etc.

For systems of a greater order the second requirement becomes ever more complex; refer to the available literature.

Example 4.3

A system with: $H(s) = \dfrac{6}{-2s - 10}$, has a denominator polynomial $N(s) = -2(s+5) = -2(s+c_0)$; because c_0 is positive, the system, as we have already seen, is stable.

Example 4.4

A system with: $H(s) = \dfrac{4}{2s^2 + 8s - 24}$, has a denominator polynomial:

$N(s) = 2(s^2 + 4s - 12) = 2(s^2 + c_1 s + c_0)$; as $c_0 < 0$ the system is unstable, as can be verified quite easily from the pole/zero plot.

Example 4.5

A system with: $H(s) = \dfrac{-3(s-3)}{s^3 + 4s^2 + 4s}$, has a denominator polynomial:

$N(s) = s^3 + 4s^2 + 4s = s^3 + c_2 s^2 + c_1 s + c_0$

Because c_1, c_2, and c_3 are all positive and $c_1 c_2 - c_0 = 4 \times 4 - 0 = 16 > 0$, this system is stable.

Example 4.6

A system with: $H(s) = 4\,\dfrac{s}{s^3 + 1}$ is unstable, as the reader can verify himself from the requirements for stability.

Let us consider:
$s^3 + 1 = (s+1)(s^2 - s + 1) = (s+1)\,(s - \tfrac{1}{2} + j\sqrt{\tfrac{3}{4}})\,(s - \tfrac{1}{2} - j\sqrt{\tfrac{3}{4}})$.

Then the result can also be verified from the position of the poles in the s-plane.

4.3 LINES OF ABSOLUTE AND RELATIVE DAMPING IN THE s-PLANE

To assess the utility of a system, whether controlled or not, very often the step response of such a system is measured. Specific demands can be made upon the response, relating to the static and dynamic behaviour of the system. The usual criteria are: the dc gain, the settling time, the overshoot, and the peak time. These criteria are determined as follows:

- The dc gain is the ratio between the steady change of the output signal and the amplitude of the input signal.
- The settling time is the time that elapses, because of a step disturbance in the input signal, before the output signal stays within a specific band around the final value of this output signal.

    If we take $\pm 2\%$ of this final value for this band, we call this time the settling time (2%) or $t_s(2\%)$, also known as the indication time. For $\pm 5\%$ we speak of settling time (5%) or $t_s(5\%)$, also known as response time.

    When the system is not linear in behaviour, for example, because of saturation, the value of the step disturbance will influence the value of the settling time.
- The overshoot is the maximum value of the positive difference between the response and the final value of this response. Usually we express the overshoot as a percentage of the final value of the response.
- The peak time is the time that elapses from the start of the disturbance to the occurrence of overshoot.

Fig. 4.3 illustrates these criteria:

dc gain B/A, the settling time (5%), the settling-time (2%), the overshoot D, and the peak-time t_p.

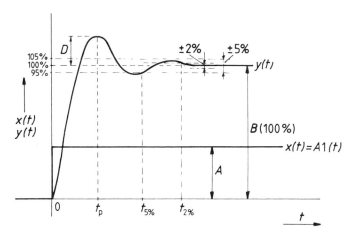

Fig. 4.3 — Criteria in the step response.

Regarding the dynamic behaviour of a system we mostly use the response or indication time and overshoot criteria. To both criteria we attach maximum values to which the system must perform. How these criteria will be applied to the position of the poles of the system in the s-plane, is demonstrated by means of a second order system with a step input signal.

We start from a stable system with two complex conjugated poles, viz., $s_1 = \lambda_1 + j\omega_1$ and $s_2 = \lambda_1 - j\omega_1$, and the pole/zero gain is indicated by $K_{pn} = K_1$. The input $x(t)$ is a step with an amplitude of M. Fig. 4.4a shows the pole/zero plot of the system, and Fig. 4.4b shows the pole/zero plot of the step response $Y(s)$.

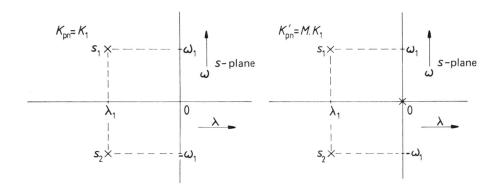

Fig. 4.4 — Pole/zero plot of $H(s)$ and $Y(s)$.

From Fig. 4.4b we obtain for $y(t)$:

$$y(t) = M \cdot K_1 \{A_1 \, e^{0t} + A_2 \, e^{(\lambda_1 + j\omega_1)t} + A_3 \, e^{(\lambda_1 - j\omega_1)t}\} \tag{4.7}$$

with $\quad A_1 = \dfrac{1}{(-\lambda_1 - j\omega_1)(-\lambda_1 + j\omega_1)} = \dfrac{1}{\lambda_1^2 + \omega_1^2}$

$$A_2 = \dfrac{1}{2j\omega_1 \cdot (\lambda_1 + j\omega_1)} = \dfrac{-j\lambda_1 - \omega_1}{2\omega_1(\lambda_1^2 + \omega_1^2)}$$

$$A_3 = \dfrac{j\lambda_1 - \omega_1}{2\omega_1(\lambda_1^2 + \omega_1^2)}$$

From (4.7) we obtain:

$$y(t) = \dfrac{MK_1}{\lambda_1^2 + \omega_1^2} \left[1 - \dfrac{e^{\lambda_1 t}}{2\omega_1} \left\{ \omega_1(e^{j\omega_1 t} + e^{-j\omega_1 t}) + j\lambda_1(e^{j\omega_1 t} - e^{-j\omega_1 t}) \right\} \right]$$

$$= \dfrac{MK_1}{\lambda_1^2 + \omega_1^2} \left[1 - e^{\lambda_1 t} \left\{ \cos \omega_1 t - \dfrac{\lambda_1}{\omega_1} \sin \omega_1 t \right\} \right] \tag{4.8}$$

To gain an insight into $y(t)$ the cosinus and sinus term are combined as follows:

Suppose:

$$\tan \varphi = \dfrac{\lambda_1}{\omega_1} \tag{4.9}$$

then:

$$y(t) = \dfrac{MK_1}{\lambda_1^2 + \omega_1^2} \left[1 - \dfrac{e^{\lambda_1 t}}{\cos \varphi} \left\{ \cos \omega_1 t \cos \varphi + \sin \omega_1 t \sin \varphi \right\} \right]$$

$$= \dfrac{MK_1}{\lambda_1^2 + \omega_1^2} \left[1 - \dfrac{e^{\lambda_1 t}}{\cos \varphi} \cdot \cos(\omega_1 t + \varphi) \right] \tag{4.10}$$

In connection with (4.9) we have:

$$\cos \varphi = \dfrac{\omega_1}{\sqrt{\lambda^2 + \omega^2}} \tag{4.11}$$

Sec. 4.3] Lines of absolute and relative damping in the s-plane

from which we obtain:

$$y(t) = \frac{MK_1}{\lambda_1^2 + \omega_1^2} \left[1 - e^{\lambda_1 t} \cdot \sqrt{1 + \frac{\lambda_1^2}{\omega_1^2}} \cdot \cos(\omega_1 t + \varphi) \right] \qquad (4.12)$$

In Fig. 4.5 the response of (4.12) is shown. The response consists of a decreasing transient phenomenon and a permanent part as follows:

$$y(t) = \underbrace{\frac{MK_1}{\lambda_1^2 + \omega_1^2}}_{\text{permanent part}} - \underbrace{\frac{MK_1}{\omega_1 \sqrt{\lambda_1^2 + \omega_1^2}} e^{\lambda_1 t} \cos(\omega_1 t + \phi)}_{\text{decreasing transient phenomenon}} \qquad (4.13)$$

Notice that for $t = 0$, we have, as might be expected, indeed $y(t) = 0$; for $t \to \infty$ we have $y(t) = \frac{MK_1}{\lambda_1^2 + \omega_1^2}$, and the dc gain is $\frac{K_1}{\lambda_1^2 + \omega_1^2}$. The amplitude of the step is M.

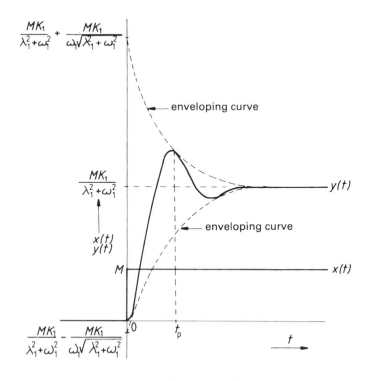

Fig. 4.5 — Step-response of the second order system.

The transient occurs within two e-functions around the final value of the response.

In practice, for $t_s(2\%)$ and $t_s(5\%)$ respectively, we often substitute the time in which the value of $e^{\lambda_1 t}$ has decreased to $\frac{1}{50}$ and $\frac{1}{20}$, by which is meant the time that passes before the transient is less than 2% or 5% of the final value of the response.

The difference of the 'correct' $t'_s(2\%)$ or $t'_s(5\%)$ and $t_s(2\%)$ or $t_s(5\%)$ respectively related to these enveloping e-functions is always such that $t_s(5\%) \geq t'_s(5\%)$ and $t_s(2\%) \geq t'_s(2\%)$ respectively; if this practical requirement is respected, the correct settling time requirement will be met.

From expression (4.13) it is clear that the enveloping e-functions do not begin at +100% and −100% around the final-value of the step response, although this was what we started from when calculating the settling time.

In fact a correction must be implemented. This correction will be proportionally smaller when λ/ω gets smaller. For $\lambda/\omega = 1$ the error is about 10%, and when $\lambda/\omega = \frac{1}{2}$, about 6%. When $\lambda/\omega = 1$, D is about 4.3%. (See also paragraph 2.2. Note that $\lambda = \omega_0 \cdot \beta$ and $\omega = \omega_0 \sqrt{1 - \beta^2}$).

If there is a requirement that $t_s(5\%) \leq t_1$, for the design of a system, then we require:

$$e^{\lambda_1 t_1} \leq \tfrac{1}{20} \tag{4.14}$$

This means that λ_1 must be negative but also that the corresponding poles must have a specific distance to the imaginary axis. Thus applies, if we suppose $e^{-3} \approx \frac{1}{20}$ and $e^{-4} \approx \frac{1}{50}$; e.g. for $t_s(5\%) \leq 3$ s, the requirement $\lambda_1 \leq -1$, and for $t_s(2\%) \leq 2$ s the requirement $\lambda_1 \leq -2$.

If a pair of conjugated poles is at a specific distance ($\lambda = \lambda_1$) from the imaginary axis then $t_s(2\%)$ and $t_s(5\%)$ are determined. A line in the s-plane for which $s = \lambda_1 + j\omega$ is called a line with constant absolute damping λ_1. A pair of conjugated poles that is to the left of the line of absolute damping $s = \lambda_1 + j\omega$ has a $t_s(5\%)$ or $t_s(2\%)$, that are shorter than is implied by: $e^{\lambda_1 t_s(5\%)} = \frac{1}{20}$, and $e^{\lambda_1 t_s(5\%)} = \frac{1}{50}$ respectively.

In Fig. 4.6 the absolute damping line $s = \lambda_1 + j\omega$ is indicated.

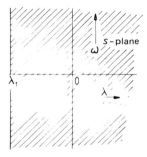

Fig. 4.6 — A line of constant absolute damping in the s-plane.

Sec. 4.3]   **Lines of absolute and relative damping in the s-plane**   63

The shading in Fig. 4.6 indicates where the poles of our second order system may not be positioned, if the absolute damping is required to be larger than $|\lambda_1|$.

For the second order system of our example:

$$t_s(5\%) = \frac{-1}{\lambda_1} \ln 20 \text{ and } t_s(2\%) = \frac{-1}{\lambda_1} \ln 50 \tag{4.15}$$

As well as assessing the behaviour of the system according to $t_s(2\%)$ and $t_s(5\%)$, a judgement of the overshoot is important. According to the definition, the overshoot occurs when the response achieves its maximum value at $t = t_p$. The determination of the extreme values of the response and the times at which these occur is obtained from (4.12):

$$\frac{d}{dt}y(t) = 0 \Rightarrow \frac{\lambda_1}{\omega_1} \cdot e^{\lambda_1 t} \cdot \cos(\omega_1 t + \varphi) - e^{\lambda_1 t} \sin(\omega_1 t + \varphi) = 0$$

or

$$\tan(\omega_1 t + \varphi) = \frac{\lambda_1}{\omega_1} = \tan\varphi$$

so:

$$\omega_1 t + \varphi = \varphi \pm k\pi$$

for $\omega_1 > 0$:

$$\omega_1 t = k\pi, \text{ so: } t = \frac{k\pi}{\omega_1}, \ k = 0, 1, \ldots \tag{4.16}$$

The first minimum for $k = 0$ is at $t = 0$; and the first maximum for $k = 1$ is at π/ω_1 s. In the same way we may determine the times of the other minima and maxima.

The peak time (this is the position of the first maximum) occurs when $t_p = \pi/\omega_1$. The value of this maximum is:

$$y(t_p) = \frac{MK_1}{\lambda_1^2 + \omega_1^2} \left[1 - \frac{e^{\lambda_1(\pi/\omega_1)}}{\omega_1} \sqrt{\lambda_1^2 + \omega_1^2} \cos(\pi + \varphi) \right]$$

$$= \frac{MK_1}{\lambda_1^2 + \omega_1^2}\left[1 + e^{\lambda_1(\pi/\omega_1)}\right] \qquad (4.17)$$

For the overshoot, expressed as a percentage of the final value of the response:

$$D = \frac{y(t_p) - y(\infty)}{y(\infty)} = e^{(\lambda_1/\omega_1)\pi} \cdot 100\% \qquad (4.18)$$

From (4.18) it appears that the overshoot is a function of the ratio λ/ω. This ratio is the relative damping. Note that ω represents the damped frequency. In the s-plane, points for which $\lambda = \omega$ is constant are lines through the origin.

Such lines are indicated as lines of constant relative damping. In Fig. 4.7 we find

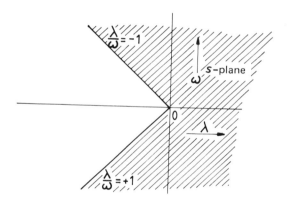

Fig. 4.7 — Lines of relative damping in the s-plane.

in the second quadrant the line of relative damping with a value of -1. This line is at an angle of 135° to the positive real axis. If one pole of a complex conjugated pair is on this line, then the overshoot (if these are the only poles of the system) equals $D = e^{-\pi} \cdot 100\% \approx 4\%$. If the overshoot is required to be smaller than 4%, the poles must be in the shaded area of the s-plane.

Remarks
1. As in the step response of a second order system with two complex conjugated poles, see e.g. (4.12), the frequency is always positive, and the relative damping lines in the second quadrant determine the overshoot that occurs in the step response.
2. Some much used values for relative damping and the overshoot that occurs are:

$$\frac{\lambda}{\omega} = -1 \quad (\beta = 0.7) \quad D \simeq 4\%$$

$$\frac{\lambda}{\omega} = -0.7 \quad (\beta = 0.6) \quad D \simeq 11\%$$

$$\frac{\lambda}{\omega} = -0.5 \quad (\beta = 0.45) \quad D \simeq 20\%$$

Example 4.7
A system is given by the differential equation:

$$2\frac{d^2y(t)}{dt^2} + 12\frac{dy(t)}{dt} + 54\,y(t) = 4\,x(t)$$

Obtain $t_s(2\%)$, $t_s(5\%)$, t_p and D.

Answer:
The transfer function is:

$$H(s) = \frac{2}{(s+3+j3\sqrt{2})\,(s+3-j3\sqrt{2})}$$

As $\lambda = -3$ and $\omega = 3\sqrt{2}$ rad/s, then: $t_s(2\%) = \tfrac{4}{3}$s, $t_s(5\%) = 1$s, $t_p = 0.74$ s and $D \simeq 11\%$.

In calculating the settling time and the overshoot we start with a second order system. In most cases it appears to be that the behaviour of the system, also in higher order systems, is determined to a high degree by two dominant complex conjugated poles. These two dominant poles are closest to the imaginary axis. The relations derived can usually also be applied for systems with more than two poles. When designing controlled systems there will be requirements as to the settling time and the overshoot of the step response, usually at the same time. This will be discussed further in Chapter 7.

4.4 NON-MINIMUM-PHASE SYSTEMS

In the previous paragraphs we concluded that systems that have a pole in the right half-plane are unstable. But what if there is a zero in the right half-plane? These systems may be stable in principle, but still form a particular group; they are called non-minimum-phase systems. Such systems occur, among others, in controlling ships and planes. Often, such a system has arisen because parallel first order systems are found here. We find an example in Fig. 4.8.

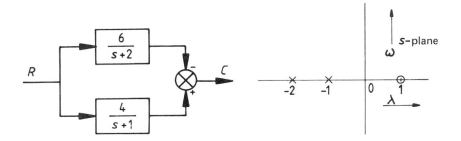

Fig. 4.8 — Non-minimum-phase system. a. Parallel first order systems: b. Pole/zero plot.

By means of a step response a particular property of these systems is explained. The transfer function $H(s)$ of the system in Fig. 4.8 is

$$H(s) = \frac{4}{s+1} - \frac{6}{s+2} = \frac{-2(s-1)}{(s+1)(s+2)} \qquad (4.19)$$

Thus the system has two poles $s = -1$ and $s = -2$ and one zero $s = +1$. A unit step input yields an output signal:

$$C(s) = \frac{-2(s-1)}{s(s+1)(s+2)} = \frac{1}{s} - \frac{4}{s+1} + \frac{3}{s+2}$$

In the time-domain this becomes:

$$c(t) = 1 - 4\,e^{-t} + 3\,e^{-2t} \qquad (4.20)$$

and

$$c(0) = 0;\ c(\infty) = 1$$

$c(t) = 0$ for $t = 0$ and $t \simeq 1.15$

$\dfrac{dc}{dt} = 0$ for $t \simeq 0.4$ (horizontal tangent)

The response is drawn in Fig. 4.9. It appears that the response first goes 'the wrong way' before going to the final value of 1. A non-minimum-phase system can often be recognized from this particular property.
In electronics a non-minimum-phase system occurs in the form of a phase shifter.

The transfer function of such a phase shifter is:

$$H(s) = \frac{-s+a}{s+a} \qquad (4.21)$$

The pole $-a$ and the zero $+a$ are symmetrical with respect to the origin. For the modulus and the argument of the transfer function we have $|H(j\omega)| = 1$ and arg $H(j\omega) = -2 \arctan \omega/a$ respectively. Please check!

Although non-minimum-phase systems are stable in themselves, instability can easily occur after feedback. An example is given in the next chapter. Moreover the systems react rather slowly because of the above step response. For these reasons we try to avoid non-minimum-phase systems.

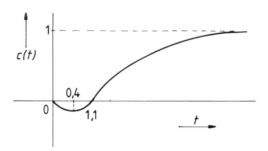

Fig. 4.9 — Step response of a non-minimum-phase system.

4.5 SYSTEMS WITH TIME DELAY

Systems with time delay (also called dead time) form a separate group, so far as analysis in the s-plane goes. Time delays occur where there are time lags present in the system. Examples may be conveyor belts, transducers that must measure velocity (e.g. flow), performing this a little later because of the physical distance of the quantity to be measured. A part of a system representing pure delay has this relation between input x and output y:

$$y(t) = x(t-T) \qquad (4.22)$$

in which T is the delay.

By means of the Laplace transform this becomes:

$$Y(s) = e^{-sT} X(s)$$

so that the transfer function $H(s)$ of a pure time delay is represented by:

$$H(s) = e^{-sT} \tag{4.23}$$

It is clear that the factor e^{-sT} cannot be expressed simply in poles and zeros. For further analysis of systems with time delay see Chapter 6.

4.6 PROBLEMS

1 Determine by means of the position of the poles in the s-plane, whether the systems below are stable or instable:

a $\quad H(s) = \dfrac{4}{s^2 - 1}$

b $\quad H(s) = \dfrac{4}{s^2 + 1}$

c $\quad H(s) = \dfrac{4s - 20}{s^3 + 6s^2 + 8s}$

d $\quad H(s) = \dfrac{1000}{s^2}$

e $\quad H(s) = \dfrac{2s + 1}{s^3 + 3s^2 + 3s + 1}$

f $\quad H(s) = \dfrac{4s + 50}{2s^3 + 8s^2 + 3s + 12}$

g $\quad H(s) = \dfrac{s - 1}{s^3 - 1}$

2. Check by means of Routh's stability criterion, whether the answers of exercise 1 are true.
3. We have the third order system of Fig. 4.10.

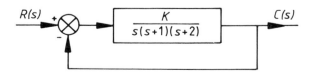

Fig. 4.10 — The system of Exercise 3.

For which values of K is this system stable?
4. In Fig. 4.11 shows a first order system.

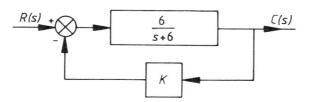

Fig. 4.11 — The system of Exercise 4.

For which values of K is this system stable?
5. We have the transfer function:

$$H(s) = \frac{100}{(s^2 + 2s + 10)(s + 10)}$$

Determine the approximate values of $t_s(2\%)$, $t_s(5\%)$, t_p and D in the step response.
6. For a second order system it is required that $t_s(5\%) \leq 1$ s and $D \leq 4\%$.
Indicate by shading in the s-plane the 'forbidden' area for the poles of this system.

7. Draw the pole/zero plot of the transfer function $H(s) = \dfrac{Y(s)}{X(s)} = \dfrac{2s - 1}{(s + 1)(5s + 1)}$.

Determine, from the pole/zero plot, the unit step response $y(t)$.
Draw this response. What is your conclusion?

5
Root loci of systems without time delay

5.1 INTRODUCTION

In the previous chapter the importance of the pole/zero notation method has been indicated. In doing that we always started from the pole/zero representation of the total, possibly controlled, system. Usually, in practice, the non-controlled system is known, and we are interested in the behaviour of the system when the system is controlled.

In this chapter it will appear that the position of the poles of a system is, among other things, dependent on the feedback method of the system and the applied gain factors.

If a system parameter (e.g.: gain) varies, then, as will appear later on, the pole positions of the closed loop system will change as well. The set of these pole positions, influenced by a variation in a system parameter, is called the root locus of the controlled system. From this root locus we may obtain valuable information on the behaviour of a system. Drawing such a root locus exactly is usually not a very simple matter. For most systems it is possible to indicate the root locus sufficiently accurately by means of construction rules for the root locus. For more complex systems and for a more accurate determination of a root locus a computer may be used. If we have the necessary software at our disposal, we can often easily determine the 'best' adjustment of controlled systems; see also Chapter 7. We shall see that for systems with delay a somewhat different approach is required. Such systems and their root loci are discussed in Chapter 6.

5.2 THE INFLUENCE OF FEEDBACK ON THE POLE/ZERO REPRESENTATION OF A SYSTEM

We begin with a negative feedback system as in the block diagram of Fig. 5.1. Here K_T is a positive frequency independent feedback factor. According to (3.11) the transfer function $H_1(s)$ may be written as:

Sec. 5.3] Influence of feedback on pole/zero representation of a system

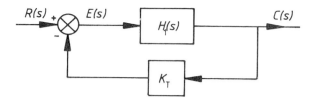

Fig. 5.1 — Block diagram of a feedback system.

$$H_1(s) = K_{pn} \frac{\prod_{i=1}^{m}(s-z_i)}{\prod_{i=1}^{n}(s-p_i)} = K_{pn} \frac{T(s)}{N(s)} \qquad (5.1)$$

For the transfer function of the controlled system, $H(s)$, we have:

$$H(s) = \frac{K_{pn} \frac{T(s)}{N(s)}}{1 + K_{pn} K_T \frac{T(s)}{N(s)}} = K_{pn} \frac{T(s)}{N(s) + K_{pn} K_T T(s)} \qquad (5.2)$$

For the feedback system it appears from formula (5.2) that:

- The zeros of the feedback system are those values of s for which the numerator of $H(s)$ is zero or:

$$T(s) = 0.$$

These appear to be the same zeros as those of the open loop system, viz. $s = z_i$, with $i = 1, 2, 3 \ldots m$.
- The poles of the feedback system are those values of s for which the denominator of expression (5.2) is zero, or:

$$N(s) = K_{pn} K_T T(s) = 0 \qquad (5.3)$$

Only for $K_{pn} \cdot K_T = 0$ follow the same poles as for the open loop system, viz. $N(s) = 0$, so $s = p_i$ with $i = 1, 2, 3 \ldots n$.

Equation (5.3) is called the root locus equation, because this equation describes how the positions of the poles shift as a function of the variation in the factor $K_{pn} K_T$.

Conclusions
1. If in a feedback system according to Fig. 5.1 the factor $K_{pn} \cdot K_T$ is varied, the position of the zeros does not change; these zeros stay the same as those of the open loop system.
2. The position of the poles of a feedback system does change if the factor $K_{pn} K_T$ is varied. The position of the poles follows from the root locus equation, and the set of all these pole positions for $0 \leqq K_{pn} K_T \leqq \infty$ is called the root locus.

In practice we choose K_T as a fixed value and frequency independent because of the sensitivity for parameter variations. Often K_T will represent the transfer function of a transducer. It has been usual to start with a unit feedback $K_T = 1$, which in principle does not matter very much, for the complete transfer function of the feedback system always has the form of expression (5.2). But, of course, to the static behaviour of the system it is important which factors are being varied.

Figs 5.2(a) and (b) indicate how a non-unity feedback system can be transformed into a system with unit feedback.

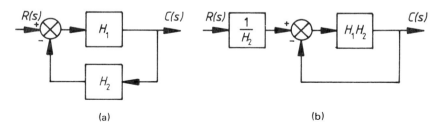

Fig. 5.2 — Transform into feedback.

By means of two simple examples the conception of the root locus will be explained.

Example 5.1
A block diagram of a feedback first order system is shown in Fig. 5.3(a)
The open loop system has one pole $s = 1/\tau$. After feedback the transfer function is:

$$H(s) = \frac{C(s)}{R(s)} = \frac{K}{\tau s + 1 + K} \tag{5.4}$$

Sec. 5.3] Influence of feedback on pole/zero representation of a system

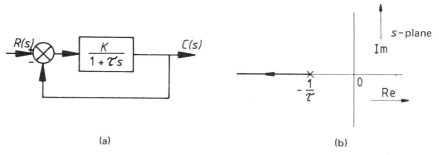

(a) (b)

Fig. 5.3 — Root locus of a feedback first order system. a. Feedback first order system. b. Root locus at variable gain.

So the closed loop system has a pole of which the position depends on K viz. $s = -\dfrac{1+K}{\tau}$. For $K = 0$ this pole appears to be equal to the one in the open loop system. The larger K becomes, the further away from the origin the pole will be on the negative real axis. The set of these poles as a function of K is the root locus. See Fig. 5.3(b).

Example 5.2
A feedback second order system is shown in Fig. 5.4.

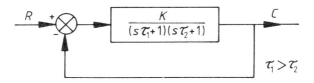

Fig. 5.4 — Feedback second order system.

The transfer function of the closed-loop system is:

$$H(s) = \frac{C(s)}{R(s)} = \frac{K}{(s\tau_1 + 1)(s\tau_2 + 1) + K} \qquad (5.5)$$

The poles comply with the following equation:

$$(s\tau_1 + 1)(s\tau_2 + 1) + K = s^2\tau_1\tau_2 + (\tau_1 + \tau_2)s + 1 + K = 0 \qquad (5.6)$$

Equation (5.6) has two real roots if:

$$D = (\tau_1 + \tau_2)^2 - 4(1 + K)\tau_1\tau_2 > 0 \text{ or } K < \frac{(\tau_1 - \tau_2)^2}{4\tau_1\tau_2}$$

If $$K = K_1 = \frac{(\tau_1 - \tau_2)^2}{4\tau_1\tau_2}$$

the system has two coinciding roots, viz.:

$$s = -\frac{1}{2}\left(\frac{1}{\tau_1} + \frac{1}{\tau_2}\right),$$

to be determined by substituting K_1 in (5.6). For larger values of K the system has two complex conjugated poles of which the real part is

$$-\frac{1}{2}\left(\frac{1}{\tau_1} + \frac{1}{\tau_2}\right).$$

With this information it is possible to draw the root locus, see Fig. 5.5.

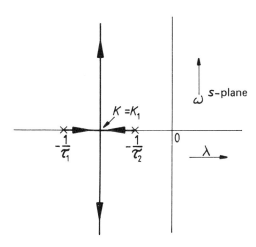

Fig. 5.5 — Root locus of a second order system.

The interpretation is clear; the value of $K = K_1$ matches $\beta = 1$ (critical damping), while $K > K_1$ and $K < K_1$ match $\beta < 1$ and $\beta > 1$ respectively. The root locus will never intersect the imaginary axis, so the poles will always stay in the left half-plane, so the system is, as we know already, stable for every positive value of K.

For these two simple examples it was easy to find the root locus, because the poles could be determined through analysis. If the transfer function of the open loop system is more complex, it is often very difficult to find analytic expressions. Therefore some general rules will be given for the construction of root loci of more complex systems.

5.3 CONSTRUCTION AND CALCULATION RULES FOR ROOT LOCI

5.3.1 Introduction

The root locus equation of formula (5.3) can always be written in the following normal form:

$$\frac{T(s)}{N(s)} = -\frac{1}{K'}, \quad \text{with } K' = K_{pn} K_T, \text{ the so called root locus gain.} \quad (5.7)$$

To derive the construction rules we shall always begin with the root locus equation in which K' is varied from zero to infinity. Later, we shall see this also for other parameters but K', a root locus may be constructed. Also, in these cases, by starting from the notation in (5.7), the same construction rules may be applied.

The most important construction rules for root loci will be discussed in sections 5.3.2 to 5.3.5. By these rules most problems may be determined with sufficient accuracy. In sections 5.3.6 to 5.3.8 some calculation rules will be discussed; these are mostly used for drawing more accurate conclusions and for carrying out calculations of root loci.

5.3.2 Starting and arrival points of root loci

The starting points of the root loci are found by substituting in (5.7) the initial value for K', so $K' = 0$. These starting points may then be determined from

$$\left. \begin{array}{l} N(s) = \prod_{i=1}^{n}(s-p_i) = 0, \text{ or} \\ s = p_i \ (i = 1, 2, 3, \ldots n) \end{array} \right\} \quad (5.8)$$

Conclusion
The starting points of the root loci are the poles of the open loop system.

Thus a root locus starts in n poles of the open loop system and consists of n root loci sections.

The arrival points of the root locus are found by substituting in (5.7) the final values for K', so $K' = \infty$. Now we have:

$$\left. \begin{array}{l} T(s) = \prod_{i=1}^{m}(s-z_i) = 0, \text{ or} \\ s = z_i \ (i = 1, 2, 3, \ldots m) \end{array} \right\} \quad (5.9)$$

Conclusion
The arrival points of the root locus are the zeros of the open loop system.

We have already seen that, in practice, always $n \geq m$, so there are, at most, equal numbers of zeros and poles. Often the number of poles is larger than the number of zeros; then there are m root locus sections from the starting points to m zeros. The remaining $n - m$ root locus sections proceed to the zeros at infinity; these zeros do not influence the system.

Conclusion
$n - m$ root locus sections proceed to infinity.

Example 5.3
The block diagram of a system is given in Fig. 5.6, in which the transfer function is

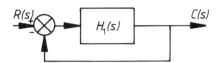

Fig. 5.6 — Unit feedback system.

given by:

$$H_1(s) = \frac{K'(s+5)}{(s+2)(s+9)} \qquad (5.10)$$

The root locus equation in the normal form is:

$$\frac{s+5}{(s+2)(s+9)} = -\frac{1}{K'} \qquad (5.11)$$

There are two starting points of the root locus, viz. $s = -2$ and $s = -9$. These are also the poles of the open loop system. There is one arrival point in $s = -5$ and one in infinity. So one of the root locus sections proceeds to infinity.

5.3.3 Asymptotes of the root locus
The $n - m$ root locus sections proceed along $n - m$ asymptotes to infinity for $K' = \infty$. These asymptotes intersect at certain angles in one point on the real axis. This intersection q is the vertex of the pole/zero plot of the open loop system. The proof for these propositions is omitted here.

The intersection of the asymptotes on the real axis is the vertex q for which applies:

$$q = \frac{\sum_{j=1}^{n} p_j - \sum_{i=1}^{m} z_i}{n - m} \quad (5.12)$$

The angles at which the asymptotes intersect the real axis in the vertex are:

$$\alpha_k = \frac{\pi + (k-1) \cdot 2\pi}{n - m}, \quad \text{with} \quad k = 1, 2, \ldots (n - m). \quad (5.13)$$

Example 5.4
The transfer function $H_1(s)$ of a system such as Fig. 5.6, is:

$$H_1(s) = 10 \frac{s + 2}{s(s + 3)(s + 5)} \quad (5.14)$$

There are three starting points viz. $s_1 = 0$, $s_2 = -3$, and $s_3 = -5$, so also three root locus sections. One of them goes to zero in $s = -2$; the others proceed to infinity, along two asymptotes. These two asymptotes intersect in the vertex of the pole/zero plot of $H_1(s)$, for which applies, according to (5.12):

$$q = \frac{0 - 3 - 5 - (-2)}{3 - 1} = -3$$

The angles at which the asymptotes intersect the real axis are:

$$\alpha_1 = \frac{\pi + 0}{2} = \frac{\pi}{2}$$

$$\alpha_2 = \frac{\pi + 2\pi}{2} = \frac{3\pi}{2}$$

5.3.4 Symmetry of the root locus

As the pole/zero plot of a system is always symmetrical with regard to the real axis (for pole and zero pairs always appear in complex conjugate pairs), we find that also the root locus itself is symmetrical with regard to the real axis. Remember that the root locus is the representation of all poles for K' from 0 to ∞. For every value of K' it is possible to determine, from the root locus, the matching pole/zero plot of the complete system.

Example 5.5
Suppose $H_1(s)$ in the system of Fig. 5.6 is given by:

$$H_1(s) = 5\frac{s+2}{s+6} \tag{5.15}$$

The one root locus section forming the system starts in $s = -6$ and proceeds to the point of arrival in $s = -2$. Because of the symmetry requirement, this root locus section can only follow the real axis. The root locus of this system for K' from $0 \to \infty$ is shown in Fig. 5.7. In the root locus sections the arrows show how the poles progress at increasing values of K'.

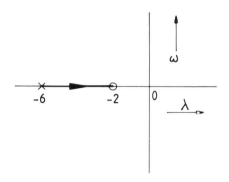

Fig. 5.7 — Root locus of the system with $H_1(s) = K'\frac{s+2}{s+6}$.

5.3.5 Parts of the root locus on the real axis
The root locus equation in the normal form of (5.7) usually represents a complex equation from which two requirements may be derived:

$$\left|\frac{T(s)}{N(s)}\right| = \left|-\frac{1}{K'}\right| = \frac{1}{K'} \tag{5.16}$$

and

$$\arg\left\{\frac{T(s)}{N(s)}\right\} = \arg\{T(s)\} - \arg\{N(s)\} = \pi \pm 2k\pi \quad \text{with } k = 0, 1, 2, \ldots \tag{5.17}$$

Each point of the s-plane that satisfies equation (5.7) also satisfies the requirements found in (5.16) and (5.17); these are all points of the root locus. The requirement in the modulus condition (5.16) does not impose any substantial limit for K' for $0 \to \infty$; however, the condition of (5.17) is called the phase condition for points belonging to the root locus.

For all points on the real axis:

$$\arg\left\{\frac{T(s)}{N(s)}\right\} = (1 \pm 2k)\pi, \text{ with } k = 0, 1, 2, \ldots \qquad (5.18)$$

All points on the real axis for which (5.17) is applicable, are points of the root locus and thus form the parts of the root locus on the real axis. The phase of some points on the real axis, and whether they belong to the root locus or not, are indicated in Fig. 5.8(a), (b), (c), and (d). Fig. 5.8(e) shows the root locus of the system.

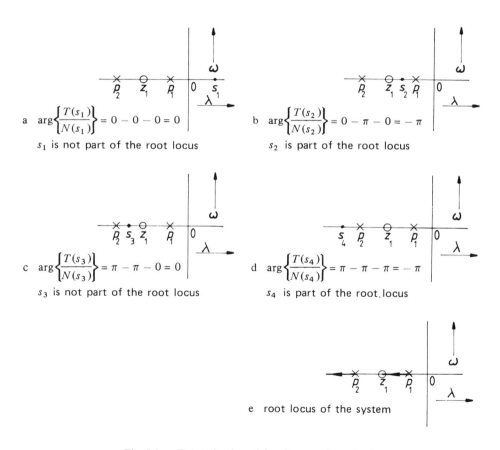

Fig. 5.8 — Determination of the phase on the real axis.

The application of the phase condition to determine the root locus sections on the real axis leads to the following conclusion.

Conclusion
All those points on the real axis for which (proceeding from right to left along this axis) the number of poles and zero passed on the real axis is odd, belong to the root locus.

Example 5.6
Consider again the system of Fig. 5.6, in which we now have for $H_1(s)$:

$$H_1(s) = K' \frac{(s+4)(s^2+10s+34)}{s(s+2)^2(s+6)} \tag{5.19}$$

Fig. 5.9 indicates which parts of the real axis belong to the root locus. Also indicated is the direction of the poles from the starting point of the root locus sections for an increasing value of K'. Note that this is only a part of the root locus, as the root locus sections between $s = 0$ and $s = -2$ have no arrival points so far. Also we can see from this example that poles and zeros which are not on the real axis, have no influence on the root locus sections on the real axis, because the total phase contribution of complex conjugate poles or zeros as to a point on the real axis equals zero.

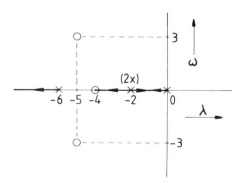

Fig. 5.9 — Partial root locus of the system with transfer function
$$H_1(s) = K' \frac{(s+4)(s^2+10s+34)}{s(s+2)^2(s+6)}.$$

5.3.6 Arrival and starting points on the real axis

From the example in Fig. 5.9 it appears that two poles influenced by increasing K' are approaching along the real axis between $s = -2$ and $s = 0$. Somewhere between these two points the two will meet and coincide. It is likely that they will then start from the real axis and proceed to the arrival points. If we also consider the example of Fig. 5.5, we shall see that the same phenomenon occurs. Two poles are coming together along the real axis and will coincide for certain values of K'. After that the poles become complex conjugate for even larger values of K'; they will then start from the real axis in the so called 'breakaway-point'.

If there are two zeros on the real axis and the part of the real axis between these zeros belongs to the root locus, then two root locus sections should arrive somewhere on this part of the real axis. From this point of arrival the two root locus sections will proceed to the finites of these sections, the zeros in question.

In a starting or arrival point the root locus equation has two or more equal roots; in such a point p_1 the root locus equation for two coinciding poles may be written as:

$$F(s) = N(s) + K' T(s) = (s - p_1)^2 \cdot G(s) = 0 \tag{5.20}$$

For the first derivative of $F(s)$, in $s = p_1$, we have:

$$\left\{ \frac{dF(s)}{ds} \right\}_{s=p_1} = 2(s - p_1) G(s) + (s - p_1)^2 \cdot \frac{dG(s)}{ds} = 0 \tag{5.21}$$

From application to the root locus equation we obtain:

$$\frac{dF(s)}{ds} = N'(s) + K' T'(s) = 0$$

or $\qquad K' = -\dfrac{N'(s)}{T'(s)} \tag{5.22}$

By this we have:

$$F(s) = N(s) - \frac{N'(s)}{T'(s)} \cdot T(s) = 0$$

and $\qquad N(s) \cdot T'(s) - N'(s) \cdot T(s) = 0 \tag{5.23}$

From the root locus equation also follows:

$$K' = -\frac{N(s)}{T(s)}, \quad \text{and}$$

$$\frac{dK'}{ds} = \frac{N(s)T'(s) - N'(s) \cdot T(s)}{T(s)^2} \tag{5.24}$$

From (5.23) and (5.24) there follows for arrival and starting points:

$$\frac{dK'}{ds} = 0 \tag{5.25}$$

or
$$\frac{dK'}{ds} = -\frac{d}{ds}\left\{\frac{N(s)}{T(s)}\right\} \tag{5.26}$$

Now also applies:

$$\frac{d}{ds}\left\{\frac{\prod_{i=1}^{n}(s-p_i)}{\prod_{i=1}^{m}(s-z_i)}\right\} = \frac{\prod_{i=1}^{m}(s-z_i)\left\{\prod_{i=2}^{n}(s-p_i) + \prod_{\substack{i=1\\i\neq 2}}^{n}(s-p_i)\ldots\right\}}{\left\{\prod_{i=1}^{m}(s-z_i)\right\}^2} +$$

$$-\frac{\prod_{i=1}^{n}(s-p_i)\left\{\prod_{i=2}^{m}(s-z_i) + \prod_{\substack{i=1\\i>2}}^{m}(s-z_i)\ldots\right\}}{\left\{\prod_{i=1}^{m}(s-z_i)\right\}^2} = 0$$

or
$$T(s) \cdot N(s)\left\{\frac{1}{s-p_1} + \frac{1}{s-p_2} + \ldots + \frac{1}{s-p_n}\right\} +$$

$$-N(s) \cdot T(s)\left\{\frac{1}{s-z_1} + \frac{1}{s-z_2} + \ldots + \frac{1}{s-z_m}\right\} = 0$$

and after dividing this by $T(s) = N(s)$ this becomes:

$$\sum_{i=1}^{n}\frac{1}{s-p_i} - \sum_{i=1}^{m}\frac{1}{s-z_i} = 0 \tag{5.27}$$

Conclusion
Arrival and starting points of root loci on the real axis may be determined by means of formula (5.25) or with the derived formula (5.27).

Example 5.7
We have a system such as that shown in Fig. 5.6, with:

$$H_1(s) = K' \frac{s+8}{(s+2)(s+6)}. \tag{5.28}$$

The sections of the real axis between $s = -2$ and $s = -6$ and to the left of $s = -8$ are part of the root locus. We determine the location of possible starting and arrival points by means of (5.27):

$$\frac{1}{s+2} + \frac{1}{s+6} - \frac{1}{s+8} = 0$$

and from which follows:

$$(s+6)(s+8) + (s+2)(s+8) - (s+2)(s+6) = 0$$

and:

$$s^2 + 16s + 52 = (s+8)^2 - (\sqrt{12})^2 = 0$$
$$\text{therefore } s_1 = -8 + \sqrt{12} \approx -4.54$$
$$s_2 = -8 - \sqrt{12} \approx -11.46$$

It is clear that s_1 is a starting point and s_2 an arrival point.
As will be proved later, in this case, for so far as the poles are complex conjugate, the complex part of the root locus is a circle with radius $\sqrt{12}$ and a centre at $s = -8$; the root locus is sketched in Fig. 5.10.

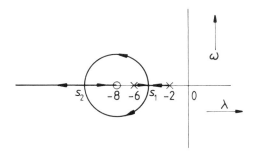

Fig. 5.10 — Root locus of the system with $H_1(s) = K' \frac{s+8}{(s+2)(s+6)}$.

If we consider the angles at which the root locus sections arrive or start, we may observe:

$$\varphi_j = \frac{180° + (j-1) \cdot 360°}{k}, \quad \text{with } j = 1, 2, 3, \ldots, k \tag{5.29}$$

Here k is the number of root locus sections that start simultaneously from one point on the real axis or arrive simultaneously at one point on the real axis.

5.3.7 The summation rule
According to the equations of (5.3) and (5.7), for the root locus we have:

$$\prod_{i=1}^{n}(s - p_i) + K' \prod_{i=1}^{m}(s - z_i) = 0 \tag{5.30}$$

Each point of the root locus satisfies this equation, and for every value of K', follows also a combination of poles of the (closed loop) system. Suppose we call the poles s_i that follow from certain values of K':

$$s_i = p_i^{\Delta} \tag{5.31}$$

Then we may assume that these poles are solutions of the equation:

$$\prod_{i=1}^{n}(s - p_i^{\Delta}) = 0 \tag{5.32}$$

Combining equations (5.30) and (5.32) gives:

$$\prod_{i=1}^{n}(s - p_i) + K' \prod_{i=1}^{m}(s - z_i) = \prod_{i=1}^{n}(s - p_i^{\Delta}) \tag{5.33}$$

Partial solution of equation (5.33) gives:

$$\left\{ s^n - \sum_{i=1}^{n} p_i \cdot s^{n-1} + \ldots + \prod_{i=1}^{n} -p_i \right\} +$$

$$+ K' \left\{ s^m - \sum_{i=1}^{m} z_i s^{m-1} + \ldots + \prod_{i=1}^{m} -z_i \right\} =$$

Sec. 5.3] Construction and calculation rules for root loci

$$= s^n - \sum_{i=1}^{n} p_i^\Delta s^{n-1} + \ldots + \prod_{i=1}^{n} -p_i^\Delta \tag{5.34}$$

On condition that $n - m \geq 2$, from (5.34), there follows the so called summation rule:

$$\sum_{i=1}^{n} p_i = \sum_{i=1}^{n} p_i^\Delta \tag{5.35}$$

Conclusion
If there are at least two poles more than there are zeros, the sum of the poles of the open loop system always equals the sum of the poles of the closed loop system.

Example 5.8
Given the system of Fig. 5.6 with:

$$H_1(s) = \frac{K'}{(s+2)(s+6)(s+8)}; \tag{5.36}$$

for certain values of K' there are two complex conjugate poles on the imaginary axis. In this case the position of the third pole can be readily determined by means of the summation rule, for there are three poles more than there are zeros. In this case

$$\sum_{i=1}^{3} p_i = \sum_{i=1}^{3} p_i^\Delta,$$

or $\quad -2 - 6 - 8 = +j\omega_1 - j\omega_1 + p_3 \Rightarrow p_3 = -16$

5.3.8 The multiplication rule
Further to equation (5.34), we also have:

$$\prod_{i=1}^{n} -p_i + K' \prod_{i=1}^{m} -z_i = \prod_{i=1}^{n} -p_i^\Delta \tag{5.37}$$

This is called the multiplication rule; there is no limit as to the number of poles with respect to the number of zeros.

The application of the multiplication rule presents the difficulty that in this formula the (unknown) factor K' is found. Now if at least one zero is in the origin, it is much easier to apply expression (5.7) because the following now applies:

$$\prod_{i=1}^{n} -p_i = \prod_{i=1}^{n} -p_i^{\Delta}$$

and so also

$$\prod_{i=1}^{n} p_i = \prod_{i=1}^{n} p_i^{\Delta} \qquad (5.38)$$

Conclusion
If at least one pole is in the origin of the *s*-plane, then the product of the poles of the closed loop system always equals the product of the poles of the open loop system.

In the pole/zero representation of physical systems these will never be just a zero in the origin, because this would imply a pure differentiator.

However, when there is no zero in the origin of the *s*-plane, we may still use formula (5.38) by starting from a transformed imaginary axis that does pass through one of the zeros. Naturally there must be a zero in the system.

Example 5.9
By means of the multiplication rule we may readily obtain the proof, already discussed in example 5.7, of the partially circular part of the root locus in question.

The above root locus is indicated again in Fig. 5.11. The imaginary axis is transformed in such a way that it passes through the zero in $s = -8$.

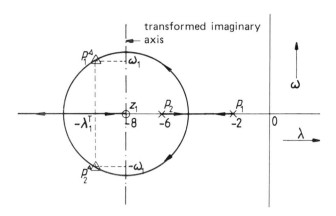

Fig. 5.11 — Proof of the circular part of the root locus.

Then we take any two complex conjugate poles on the root locus: p_1^{Δ} and p_2^{Δ}; the poles of the open-loop system are p_1 and p_2. With regard to the transformed coordinates we have:

Sec. 5.3]   **Construction and calculation rules for root loci**   87

$$p_1^T = +6, \quad p_2^T = +2$$
$$p_1^{\Delta T} = -\lambda_1^T + j\omega_1, \quad p_2^{\Delta T} = -\lambda_1^T - j\omega_1$$

becoming, according to (5.38):

$$p_1^T \cdot p_2^T = p_1^{\Delta T} \cdot p_2^{\Delta T}$$
or $\quad 6 \cdot 2 = (\lambda_1^T + j\omega_1)(-\lambda_1^T - j\omega_1) = (\lambda_1^T)^2 + \omega_1^2.$

Evidently, so far as p_1^Δ and p_2^Δ are complex conjugate, we have $(\lambda^T)^2 + \omega_1^2 = 12$, which means that this part of the root locus is a circle with its centre in the origin of the transformed coordinates and a radius equal to $\sqrt{12}$, as expected.

5.3.9 Determination of the root locus and dc gain

Often we wish to determine the value of K' from a given root locus and a certain position of the poles of the closed loop system. From expression (5.30) we have, for $s = p_1^\Delta$:

$$K' = -\frac{\prod_{i=1}^{n}(p_1^\Delta - p_i)}{\prod_{i=1}^{m}(p_1^\Delta - z_i)} \qquad (5.39)$$

As K' is always positive in the case of negative feedback, there is also a rule that may be more readily applied:

$$K' = \frac{\prod_{i=1}^{n}|p_1^\Delta - p_i|}{\prod_{i=1}^{m}|p_1^\Delta - z_i|} \qquad (5.40)$$

Conclusion

The root locus gain at a specific point of the root locus is found by dividing the product of the lengths of the vectors from the poles of the open loop system to that specific point by the product of the lengths of the vectors from the zeros to that specific point.

However, to determine the static accuracy of a controlled system, it is necessary to determine the dc gain in the loop for certain positions of the poles on the root locus.

Considering a feedback system as in Fig. 5.1, we have:

$$K_L = K' \frac{\prod_{i=1}^{m} -z_i}{\prod_{i=1}^{n} -p_i} \tag{5.41}$$

or

$$K_L = K' \frac{\prod_{i=1}^{m} |z_i|}{\prod_{i=1}^{n} |p_i|} \tag{5.42}$$

Conclusion
The loop dc gain is found by multiplying the root locus gain at a specific setting by the product of the moduli of the zeros and dividing it by the product of the moduli of the poles.

Example 5.10
A system with a block diagram as in Fig. 5.6 and with:

$$H_1(s) = \frac{K'}{(s+2)(s+8)} \tag{5.43}$$

has been adjusted so that the damping ratio is $\beta = \frac{1}{2}\sqrt{2}$.
What is the static accuracy? As the relative damping factor is $\beta = \frac{1}{2}\sqrt{2}$, the poles are at $s_1 = -5 + j5$ and $s_2 = -5 - j5$ (Fig. 5.12). For the root locus gain we have:

$$K' = \frac{l_1 \cdot l_2}{1} = \sqrt{3^2 + 5^2} \cdot \sqrt{3^2 + 5^2} = 34.$$

and for the loop dc gain:

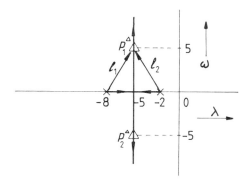

Fig. 5.12 — Root locus of the system with $H_1(s) = \dfrac{K'}{(s+2)(s+8)}$.

$$K_L = K' \frac{1}{2 \cdot 8} = \frac{34}{16} = \frac{17}{8}$$

The steady state error because of a step input change becomes:

$$E = \frac{100\%}{1 + K_L} = \frac{100\%}{1 + \frac{17}{8}} \approx 32\%.$$

5.3.10 Summary of construction and calculation rules for root loci

Assume the system of Fig. 5.13; the root locus equation is:

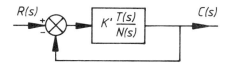

Fig. 5.13 — Block diagram of an unit-feedback system.

$$\frac{T(s)}{N(s)} = -\frac{1}{K'} \quad \text{with} \quad T(s) = \prod_{i=1}^{m}(s - z_i) \quad \text{and} \quad N(s) = \prod_{i=1}^{n}(s - p_i).$$

where z_i: zero i of the open loop system;
p_i: pole i of the open loop system;
$n \geqq m$.

Rules
1. The position of the zeros is independent of K'.
2. The position of the poles of the closed loop system follows from the root locus equation.
3. The root locus exists of n root locus sections.
4. Each root locus section starts in one pole of the open loop system.
5. m root locus sections end in the (visible) zeros; while $n - m$ root locus sections go to infinity.
6. There are $n - m$ asymptotes along which $n - m$ root locus sections proceed to infinity. All asymptotes intersect the real axis in the vertex q of the pole/zero plot at angles α_k; for q and α_k applies:

$$q = \frac{\sum_{i=1}^{n} p_i - \sum_{i=1}^{m} z_i}{n - m}, \quad \alpha_k = \frac{\pi + (k-1)2\pi}{n - m} \quad \text{with} \quad k = 1, 2, \ldots, (n - m)$$

7. The root locus is symmetrical about the real axis.
8. All those points on the real axis are part of the root locus, so that the total number of poles and zero to the right of these points is odd.
9. Possible arrival and starting points on the real axis may be determined from:

$$\frac{dK'}{ds} = 0 \quad \text{or from} \quad \sum_{i=1}^{n} \frac{1}{s - p_i} - \sum_{i=1}^{m} \frac{1}{s - z_i} = 0.$$

10. If there are two poles more than there are zeros, the sum of the poles is independent of K'.
11. If there is at least one zero in the origin, the product of the poles is independent of K'.
12. For the root locus gain at a specific point p_1^A of the root locus we have:

$$K' = \frac{\prod_{i=1}^{n} |p_1^A - p_i|}{\prod_{i=1}^{m} |p_1^A - z_i|}$$

13. For the loop dc gain at a specific point of the root locus we have:

$$K_L = K' \prod_{i=1}^{m} \frac{|z_i|}{\prod_{i=1}^{n} |p_i|}$$

14. The intersections of the root locus sections with the imaginary axis may be found by substituting $s = j\omega$ in the root locus equation. This also yields the oscillation frequency and the root locus gain when oscillating.
15. If several root locus sections start or arrive from one point on the real axis, they do this at an angle α_k, as defined by Rule 6.

5.3.11 Examples of application

In this section the root loci of a number of systems will be given; the reader may verify for himself the correctness of these root loci by means of the construction rules. Some cases will be considered in more detail.

In Fig. 5.14 we find the root loci of some systems, configured as in Fig. 5.13.

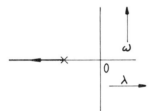

Fig. 5.14a — One pole.

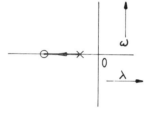

Fig. 5.14b — One pole and one zero.

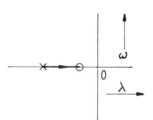

Fig. 5.14c — One pole and one zero.

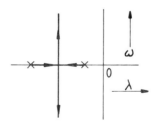

Fig. 5.14d — Two different real poles.

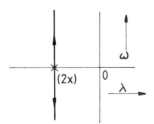

Fig. 5.14e — Two coinciding poles.

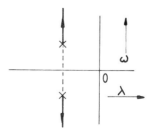

Fig. 5.14f — Two complex conjugated poles.

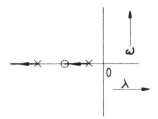

Fig. 5.14g — Two poles and one zero.

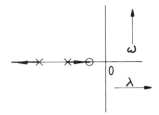

Fig. 5.14h — Two poles and one zero.

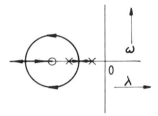

Fig. 5.14i — Two poles and one zero.

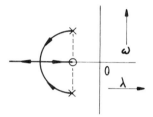

Fig. 5.14j — Two complex conjugated poles and one zero.

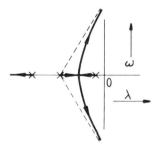

Fig. 5.14k — Three poles.

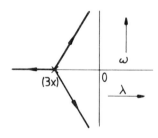

Fig. 5.14l — Three coinciding poles.

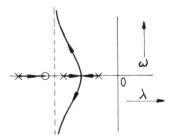

Fig. 5.14m — Three poles and one zero.

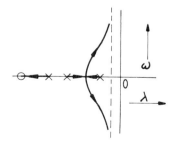

Fig. 5.14n — Three poles and one zero.

Example 5.11

A system has three time constants of $1, \frac{1}{4}$, and $\frac{1}{10}$ second respectively, and a dc gain of 10. This system is provided with a negative unit feedback. A gain factor $K(K>0)$, that may be varied, will be included in the forward path of the loop.

Draw as accurately as possible the root locus for the variation of K from $0 \to \infty$, and also determine the value of K at which this system becomes unstable. Then obtain also the oscillation frequency.

Answer
The block diagram of the system is shown in Fig. 5.15.

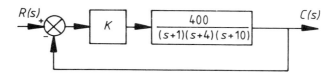

Fig. 5.15 — Block diagram of the system of Example 5.11.

The root locus equation is:

$$\frac{400}{(s+1)(s+4)(s+10)} = -\frac{1}{K}.$$

The root locus has these features:

There are three root locus sections and also three asymptotes; these intersect at the point $q = \dfrac{-1-4-10}{3} = -5$ on the real axis at the following angles: $\alpha_1 = 60°$, $\alpha_2 = 180°$, and $\alpha_3 = 300° \triangleq -60°$ with the real axis. The sections of the real axis between $s = -1$ and $s = -4$ and to the left $s = -10$ are part of the root locus. The starting and arrival points follow from $\dfrac{1}{s+1} + \dfrac{1}{s+4} + \dfrac{1}{s+10} = 0$. Now we find the condition $s^2 + 10s + 18 = 0$, or $s_1 = -2.35$ and $s_2 = -7.65$. Point s_1 satisfies and is a starting point; Point s_2 belongs to a negative value of K.

With these results the root locus becomes as shown in Fig. 5.16. The system becomes unstable as soon as K' is increased, so that two poles of the closed loop system are on the imaginary axis ($p_1^\triangle$ and $p_2^\triangle$).

By means of the summation rule $(n-m \geq 2)$ we obtain for the third pole: $p_3^\Delta = -15$. For the value of $K = K_{osc}$, $s = -15$ is apparently a solution of the root locus equation. Substituting $s = -15$ in the root locus equation yields:

$$\frac{400}{(-15+1)(-15+4)(-15+10)} = \frac{1}{K_{osc}}, \text{ or: } K_{ose} = 1.925$$

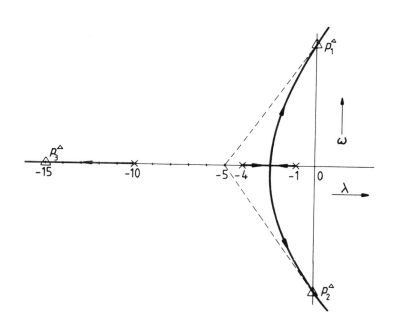

Fig. 5.16 — Root locus of the system in Example 5.11.

The generated frequency can be found by substituting this value of K_{osc} in the root locus equation $s = p_1^\Delta = j\omega_{osc}$ or $s = p_2^\Delta = -j\omega_{osc}$, for these are also solutions of the root locus equation.

We substitute $s = j\omega_{osc}$ and find:

$$\frac{400}{(j\omega_{osc}+1)(j\omega_{osc}+4)(j\omega_{osc}+10)} = -\frac{1}{1.925} \tag{5.44}$$

or $\quad (j\omega_{osc})^3 + 15(j\omega_{osc})^2 + 54j\omega_{osc} + 40 = -770$

Therefore:

$$-j\omega_{osc}^3 + 54j\omega_{osc} = 0$$
and $\quad -15j\omega_{osc}^2 + 40 = -770$

From both conditions follows: $\omega_{osc} = \sqrt{54} \approx 7.35$ rad/s.

Note:
As we obtain two equations from (5.44) (the real and the imaginary part of this equation), we may also obtain K_{osc}. Please check!

Example 5.12
We have a system as shown in Fig. 5.17.

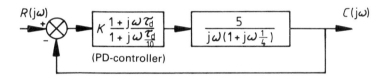

Fig. 5.17 — Block diagram of Example 5.12 in the ω-domain.

Question
Determine τ_d in such a way that the oscillatory transient phenomenon in the response has, because of a step input signal, an absolute damping of $\lambda = -20$ and becomes independent of K. Also obtain the value of K, for which in that case $\beta = \frac{1}{2}\sqrt{2}$.

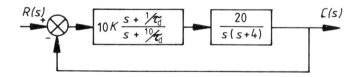

Fig. 5.18 — Block diagram of Example 5.12 in the s-domain.

Answer
The block diagram in the s-domain is sketched in Fig. 5.18. The root locus equation becomes:

$$\frac{200\left(s + \frac{1}{\tau_d}\right)}{s\left(s + \frac{10}{\tau_d}\right)(s+4)} = -\frac{1}{K} \tag{5.45}$$

If the absolute damping is to become independent of K, the root locus must proceed parallel to the imaginary axis. This is the case when there are only two poles; see also Example 5.2. Now choose τ_d so that the zero compensates for the pole $s = -4$, so $\tau_d = \frac{1}{4}$ second, then the root locus equation will be:

$$\frac{200}{s(s+40)} = -\frac{1}{K} \tag{5.46}$$

The root locus will be as in Fig. 5.19.

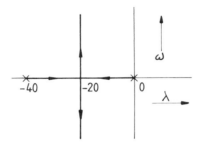

Fig. 5.19 — Root locus of the system in Example 5.12.

For $\tau_d = \frac{1}{4}$ s, the compensation in question occurs, and if the poles are complex conjugate, the absolute damping becomes constant and equal to 20.

For $\beta = \frac{1}{2}\sqrt{2}$, $s_1 = -20 + j20$ and $s_2 = -20 - j20$; see also Example 5.10. If we substitute the pole $s_1 = -20 + j20$ in expression (5.46), then K may be determined:

$$\frac{200}{(-20+j20)(-20+j20+40)} = -\frac{1}{K} \Rightarrow K = 4.$$

Example 5.13
Given the non-minimum phase system with transfer function $H(s)$:

$$H(s) = K\frac{s-1}{(s+1)(s+2)} \tag{5.47}$$

Question
The root locus of the unity feedback system for variable gain $0 < K < \infty$. Determine the values of K at which the system is unstable.

Answer
The block diagram of the control system is shown in Fig. 5.20.

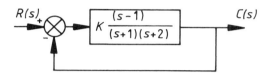

Fig. 5.20 — Feedback non-minimum phase system.

In Fig. 5.21 the root locus in question is sketched, which follows readily from the construction rules.

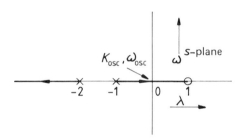

Fig. 5.21 — Root locus of the system in Fig. 5.20.

The values of K for which the system is unstable, may be found by substituting $s = j\omega$ in the root locus equation:

$$(s+1)(s+2) + K(s-1) = 0 \qquad (5.48)$$

or: $\quad s^2 + 3s + 2 + Ks - K = 0$

Substituting of $s = j\omega$ gives:

$$-\omega^2 + 3j\omega + 2 + jK\omega - K = 0$$

from which follows:

$$\begin{cases} -\omega^2 + 2 - K = 0 \\ 3\omega + K\omega = 0 \end{cases} \quad (5.49)$$

From the root locus and (5.49) it follows that $K_{osc} = 2$ and $\omega_{osc} = 0(!)$. For values of K larger than or equal to 2 the output of the system will 'drift away'.

5.4 ROOT LOCI FOR NEGATIVE VALUES OF K

Although it seems illogical at first to consider root loci for negative values of K, such a situation may occur, for example when, because of a parameter variation, a negative feedback system transforms into a positive feedback system. In this case the root locus equation is:

$$-KG(s) + 1 = 0 \quad (5.50)$$

or $\quad G(s) = +\dfrac{1}{K} \quad (5.51)$

Here we see that the phase condition changes, viz.: $\arg[G(s)] = 0$ (instead of $-\pi$), while the modulus condition remains unchanged. The consequences for the construction of the root locus concern only those rules that refer to the phase condition. The following construction rules, see also section 5.3.10, will therefore change:

- Rule 6: Concerning the direction α_k of the asymptotes:

$$\alpha_k = \frac{(k-1)2\pi}{n-m} \quad \text{with} \quad k = 1, 2, \ldots, (n-m)$$

- Rule 8: All those points on the real axis are part of the root locus for which the total number of poles and zeros to the right of these points is an *even* number.

Example 5.14
We have the non-minimum phase system with transfer function $H(s)$:

$$H(s) = -K\frac{s-1}{s(s+2)} \quad (5.52)$$

Question
The root locus of the feedback system for variable gain is $0 < K < 0$. Determine the values of K for which the system is unstable.

Answer
The block diagram of the system is shown in Fig. 5.22.

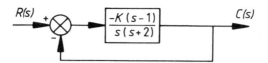

Fig. 5.22 — Feedback non-minimum phase system.

Because for the root locus equation,

$$\frac{s-1}{s(s+2)} = +\frac{1}{K} \qquad (5.53)$$

so all points of the root locus must satisfy the phase condition 0(!). The root locus sections on the real axis will be between the poles -2 and 0, and from the zero to infinity. The root locus is shown in Fig. 5.23.

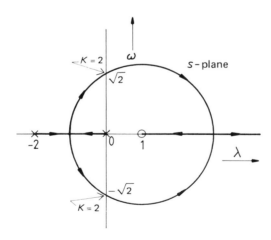

Fig. 5.23 — Root locus of the non-minimum phase system according to Fig. 5.22 for $0 < K < \infty$

If we substitute $s = j\omega$ in the root locus equation, then $K = 2$. So the system is unstable for $K \geq 2$. The starting and arrival poiints appear to be $1 - \sqrt{3}$ and $1 + \sqrt{3}$ respectively. The complex part of the root locus is a circle centred at $(1,0)$ and with a radius of $\sqrt{3}$.

5.5 COMPUTER AIDED CALCULATION AND PLOTTING OF ROOT LOCI

Except that root loci may be determined (often only approximately) by means of the above construction rules, it is also possible to calculate numerical points of the root locus. Therefore the root locus may be written as:

$$\prod_{i=1}^{n} (s - p_i) + K' \prod_{i=1}^{m} (s - z_i) = 0. \tag{5.54}$$

If in this equation the system parameters are put in in the form of poles and zeros, we shall have an equation in s of the nth degree, of which for each value of K' the roots (= poles of the closed loop system) may be determined. If we increase K' by small steps and determine the roots for each value of K', these results may be plotted in the s-plane. In this way, the root locus of the system will develop, when we have obtained sufficient solutions.

A computer will readily solve the above problem. Fig. 5.24 is a flow chart in which we see the set-up of the necessary computer progranm. First the system parameters and some other data needed in the program are put in. Then the general root locus equation is adapted for the root locus equation of the system in question, by substituting the input values of the poles and zeros. Usually the number of poles and zeros is limited; in the example the maximum number is set at 5 for both of them. After substituting the initial value of K', i.e. *START K*, the first roots of the root locus equation may be determined by an iteration process.

For this purpose several methods are known in numerical mathematics, such as Regula Falsi, Newton Raphson, and Bairstov. The determined roots (the poles of the system for a certain value of K') may then be plotted.

The number of times that a combination of poles has been determined (NK) will then be increased by one, and a test will be carried out as to whether sufficient pole combinations have been calculated. If not, K' will be increased by a step *DELTA K*. In practice this step *DELTA K* is not constant, but is each time, e.g. 10% (*EXP K* = 1.1) larger. Often, a better distribution of the plotted pole combinations will arise. By this new value of K' new poles will be calculated. As soon as $NK >$ *NKMAX*, sufficient points of the root locus have been obtained, and the program will stop.

In the example only the poles are being calculated and plotted. Often it will be possible to extend such a program fairly easily, so that the use of a computer will be even more justified. For instance, it is possible to indicate poles on the root locus for a specific value of K' or to introduce some modifications quickly in order to compare root loci.

Fig. 5.25. is an example of a possible computer output. In Fig. 5.25(a) a root locus has been plotted as in the configuration of Fig. 5.6 with:

$$H_1(s) = K \frac{s+2}{s(s+1)} \tag{5.55}$$

The step value for the increase of K was 0.05 and the number of steps was 400.

Sec. 5.5] Computer aided calculation and plotting of root loci

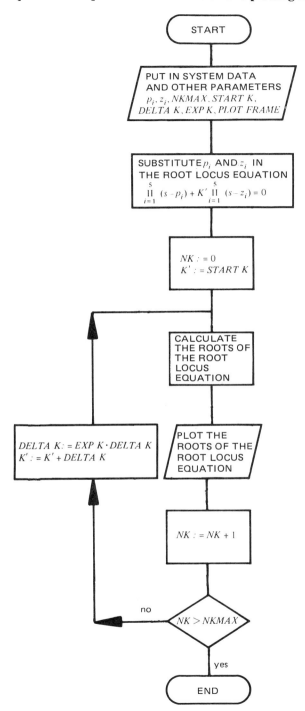

Fig. 5.24 — Flow chart for calculating and plotting a root locus.

Also the starting and arrival points have been indicated in the plot. In Fig. 5.25(b) only the step responses of $K = 0.2$; $K = 3$ and $K = 6$ have been plotted.

It will be clear that, in this way, there will be a ready understanding of the relationship between pole positions on the one hand and the dynamic properties of the system on the other.

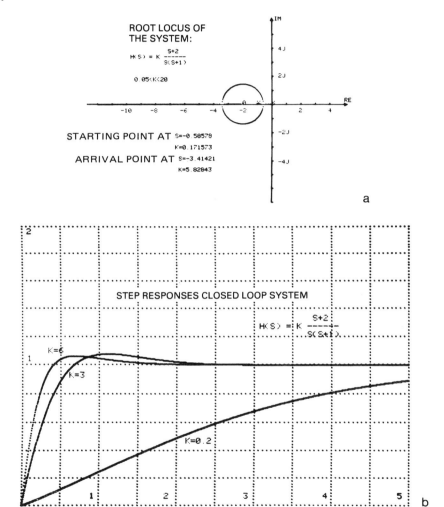

Fig. 5.25 — Computer plot of root locus and step responses.

5.6 PROBLEMS

1. Draw the root loci for $0 \leq K < \infty$ of the systems configurated as in Fig. 5.6, for the transfer function $H_1(s)$:

   a $H_1(s) = K(s+1)$

b $H_1(s) = \dfrac{K}{s}$

c $H_1(s) = \dfrac{K}{s^2}$

d $H_1(s) = \dfrac{K(s+4)}{(s+8)^2}$

e $H_1(s) = \dfrac{K(s+5)}{s^2+10s+50}$

f $H_1(s) = \dfrac{80K}{(s+4)^2}$

g $H_1(s) = \dfrac{4K(s+5)^2}{s(s+10)}$

h $H_1(s) = \dfrac{K(s+3)}{s^2}$

i $H_1(s) = \dfrac{K(s+6)}{s^2(s+4)}$

j $H_1(s) = \dfrac{K}{(s+5)^4}$

k $H_1(s) = \dfrac{K}{(s+1)(s+2)(s+3)(s+4)}$

l $H_1(s) = \dfrac{K(s+6)}{s(s^2+6s+8)}$

m $H_1(s) = \dfrac{K(s+4)}{s(s^2+8s+12)}$

n $H_1(s) = \dfrac{K(s+3)}{s(s^2+13s+40)}$

o $H_1(s) = \dfrac{Ks}{(s+3)(s^2+13s+40)}$

p $H_1(s) = \dfrac{K(s-2)}{(s+1)(s+2)(s+5)}$

2. A process is represented by the block diagram of Fig. 5.26 and the following differential equation:

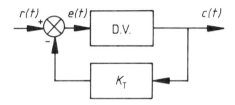

Fig. 5.26 — Block diagram of Problem 2.

$$\text{D.V.:} \quad \frac{d^3c(t)}{dt^3} + 10\frac{d^2c(t)}{dt^2} + 24\frac{dc(t)}{dt} = 4\frac{de(t)}{dt} + Ae(t)$$

Questions:
a. Draw the root locus for a variation of K_T of $0 \to \infty$, if $A = 48$.
b. Determine for which value of K_T this system becomes unstable when $A = 48$, and also determine the oscillation frequency.
c. For which maximum value of A this system cannot become unstable if $K_T > 0$?

3. A system configured as in Fig. 5.26 has, in the forward path three first order systems in cascade, and a time constant of $\frac{1}{20}$ s each. The total dc gain of these systems is 2.

Question:
a. Draw the root locus of this system for K_T from $0 \to \infty$.
b. For which value of K_T will this system become unstable, and what will be the oscillation frequency?
c. For a specific value of K_T one of the poles is at $s = -30$. Determine this value of K_T, and also calculate the position of the other poles.

4. In the system of Fig. 5.18, the control actions are given by:

$$H_R(s) = aK\frac{s + \dfrac{1}{\tau_d}}{s + \dfrac{a}{\tau_d}}.$$

Determine a and τ_d so that the absolute damping will equal $\lambda = -30$ and be independent of K, and β will equal $\frac{1}{2}\sqrt{2}$. Also determine the value of K.

5. A system has the following open loop transfer function:

$$H(s) = \frac{K(s+2)^2}{s^3(s+8)}$$

Determine the root locus of the unit feedback system for $0 < K < \infty$. What strikes you regarding the stability of the system?

6

Root loci of systems with time delay

6.1 INTRODUCTION

In many processes time delays often arise owing to the distance between the transducer (measurement of the physical parameter) and the process itself. In the literature these time delays are also called as dead time, transportation lag or delay time. To describe the influence of such a time delay for root locus purposes, it must in principle be written in the form of a combination of poles and zeros. A time delay with the transfer function:

$$H_d(s) = e^{-sT_d} \tag{6.1}$$

may be written by the mathematical limit expression

$$\lim_{n \to \infty} \left(\frac{x}{n} + 1\right)^n = e^x \tag{6.2}$$

or:

$$H_d(s) = \lim_{n \to \infty} \frac{1}{\left(1 + \frac{T_d}{n}s\right)^n} = \lim_{n \to \infty} \frac{\left(\frac{n}{T_d}\right)^n}{\left(s + \frac{n}{T_d}\right)^n} \tag{6.3}$$

According to (6.2) it is possible to represent the delay by an infinitely large poles and zeros gain and an infinite number of poles in infinity. Obviously, this consideration will not lead directly to a practical determination of the influence of a delay on the root locus of the total system. Other, more or less feasible, approximations of a delay will be discussed in section 6.4.

To determine the root locus of a system with time delay, we often use the phase lines in the s-plane.

From the path of root loci of systems with delay we may readily understand the conclusion drawn from (6.3) because, as we shall see, such root loci consist of an infinite number of root locus sections. Along these sections an infinite number of poles comes from infinity for an increasing value of the root locus gain K.

6.2 PHASE-LINES IN THE s-DOMAIN

The s-plane is a complex plane, by which a modulus and an argument from the vector to a point in the s-plane may be determined.

A phase or argument line may now be described as follows:

A phase line is a (straight) line in the s-plane, on which we find all points for which a function of s has a specific argument.

Example 6.1
Assume that $H(s) = s$; for $s = 2$ the phase of $H(s)$ equals 0, for $s = 2 + j2$ the phase is $+45°$; and for $s = j2$ the same phase equals $+90°$, etc.

All points in the s-plane for which $\omega = 0$ are one the phase line 0 or on the phase line $+180°$. All points for which $\lambda = 0$ are on the phase line $+90°$, on $+270°$, or also on $-90°$. In this way we may draw phase lines in the s-plane for each argument of $H(s)$.

Phase lines of a zero
If we assume $H(s) = s - z$, then the argument for $H(s)$ is:

$$\arg \{H(s)\} = \arg \{s - z_i\} \tag{6.4}$$

The vector $(s_1 - z_i)$ points from the zero z_i to $s = s_1$. The argument of $(s_1 - z_i)$ equals the angle of the vector $(s_1 - z_i)$ regarding the positive real axis. For each point s_i on the vector $(s_1 - z_i)$ or on the extension of it, $H(s) = (s_i - z_i)$ has the same argument; so each-straight line departing from z_i is a phase line.

In Fig. 6.1 there are some phase lines for the zero in $s = -5$.

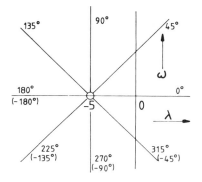

Fig. 6.1 — Some phase lines of zero in $s = -5$.

Phase lines of a pole

For $H(s) = \dfrac{1}{s - p_i}$:

$$\arg \{H(s)\} = - \arg \{s - p_i\} \tag{6.5}$$

Here the same consideration applies as for the phase lines of a zero, with due observance of the minus. Fig. 6.2 shows some phase lines of a pole in $s = -5$.

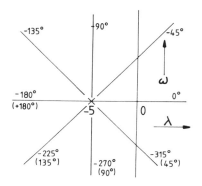

Fig. 6.2 — Some phase lines of a pole in $s = -5$.

Phase lines of a delay
For a delay:

$$H_d(s) = e^{-sT_d} = e^{-(\lambda + j\omega)T_d}. \tag{6.6}$$

The argument is:

$$\arg H_d(s) = \arg e^{-(\lambda + j\omega)T_d} = -\omega T_d \text{ (radians)} \tag{6.7}$$

This means that if ω is constant, and the argument of $H_d(s)$ is constant, so every line parallel to the λ-axis in the s-plane is a phase line of the delay.

Fig. 6.3 shows some phase lines for $T_d = \dfrac{\pi}{4} \approx 0.785 \, \text{s}$.

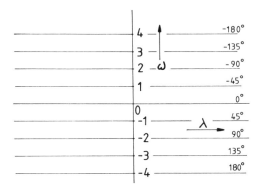

Fig. 6.3 — Some phase lines for the time delay $T_d = \dfrac{\pi}{4}$ s.

6.3 ROOT LOCUS CONSTRUCTION BY MEANS OF PHASE LINES

In Chapter 5 we have seen that there is a necessary and sufficient condition for points of the root locus:

$$\arg \frac{T(s)}{N(s)} = \pi \pm 2k\pi, \text{ with } k = 0, 1, 2, \ldots \tag{6.8}$$

This means that all points on the phase lines of $T(s)/N(s)$, and for which (6.8) applies, are also points of the root locus.

Example 6.2
Let us assume a system as in Fig. 6.4 with:

$$\mathbf{H_1(s)} = \frac{K'}{s+3}, \quad K' > 0. \tag{6.9}$$

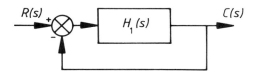

Fig. 6.4 — Configuration of the system in Example 6.2.

In Fig. 6.5 there are some phase lines of $H_1(s)$. The phase line $+180°$ (or $-180°$) is the root locus of this system for $0 < K < \infty$.

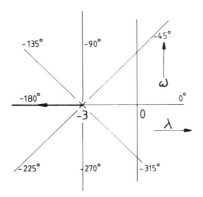

Fig. 6.5 — Some phase lines and the root locus of Example 6.2.

Example 6.3
We have the system as in Fig. 6.4 with:

$$H_1(s) = \frac{K'}{(s+2)(s+6)} \tag{6.10}$$

In Fig. 6.6 some phase lines are shown both for the pole $s = -2$ and for the pole $s = -6$. All those points in the s-plane for which the sum of the arguments of the pole at $s = -2$ and of the pole at $s = -6$, equal $\pi \pm 2k\pi$ are part of the root locus. In this way the root locus of Fig. 6.6 develops, which naturally yields the same representation as would have been found by utilizing the construction rules of Chapter 5.

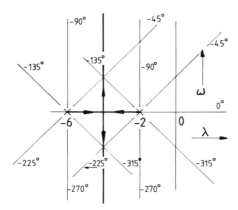

Fig. 6.6 — Some phase lines and the root locus of Example 6.3.

In both previous examples no time delay had been included in the system. In the next examples we shall discuss root loci of systems that do have a time delay.

Sec. 6.3]   **Root locus construction by means of phase lines**   111

Example 6.4
Given a system according to the configuration of Fig. 6.4:

$$H_1(s) = K'e^{-s\frac{\pi}{4}} \tag{6.11}$$

It will be clear that all those phase lines for which

$$\arg H_1(s) = -\omega\frac{\pi}{4} = 180° \pm k \cdot 360° \tag{6.12}$$

are also parts of the root loci of this system. The root locus is shown in Fig. 6.7. So the

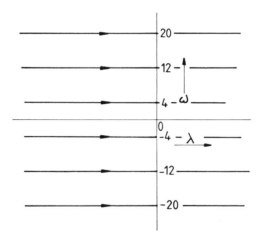

Fig. 6.7. — Root locus of a time delay with $T_d = \frac{\pi}{4}$ s.

root locus consists of those phase lines, for which

$\omega = \pm\ 4$ rad/s,

$\omega = \pm 12$ rad/s,

$\omega = \pm 20$ rad/s, etc.

The direction of the poles on the root locus, with increasing root locus gain, is indicated by arrows and from left to right; for the root locus sections depart from the (infinitely many) poles in $-\infty$. When for one specific value of the root locus gain $K' = K_1$, all poles of the closed loop system are on a vertical line through $\lambda = \lambda_i$.

We obtain the value of λ_1 from:

$$\left| e^{-s(\pi/4)} \right| = \left| -\frac{1}{K_1} \right| \tag{6.13}$$

or: $\quad e^{-\lambda_1 \pi/4} = \frac{1}{K_1} \Rightarrow \lambda_1 = \frac{4}{\pi} \ln K_1 \tag{6.14}$

For $K < 1$, and thus an absolute damping $\lambda < 0$, all poles of the closed loop system are to the left of the imaginary axis; for $K' = 1$ all poles are on the imaginary axis, and for $K' > 1$ all poles are to the right of the imaginary axis. In the first case ($K' < 1$) the system is stable; for the latter systems ($K' \geqq 1$) it is unstable.

In all cases the response to, for example, a step input signal consists of the summation of an infinite number of sinus and or cosinus terms. When $K' < 1$ the output signal contains an e-function with a negative exponent; when $K' = 1$ this e-function has a value of one, and when $K' > 1$ e-function has a positive exponent. Fig. 6.8 shows the step response of this system for $K' = \frac{1}{2}$, $K' = 1$, and $K' = 1\frac{1}{2}$.

The rectangular form of these step responses is caused by the summation of infinitely many sinus and/or cosinus terms. Here we must mention the relation between this method of signal and system description and the description by means of a Fourier expansion. The decreasing, constant, or increasing nature of the response is caused by the e-function by which all terms in the response are being multiplied.

The reader may easily find an explanation for the responses in Fig. 6.8, by assuming a step input signal and the suspense time of the signal in the delay, the system is assumed to be in a pipeline.

Example 6.5

Let us assume a system as in Fig. 6.4:

$$H_1(s) = \frac{K' \cdot e^{-s}}{s+1} \tag{6.15}$$

The time delay in this system is 1 second. Also, we have a first order system with a time constant of 1 second. For both the delays — the horizontal lines — and the pole at $s = -1$ we show some phase lines in Fig. 6.9.

All those points for which the phase of the delay and of the pole are together $\pi \pm 2k\pi$, are points of the root locus. The points on the real axis to the left of $s = -1$ satisfy this conclusion. Also the points P and Q satisfy the above requirement. For points that are far to the right, the phase lines are π and $-\pi$ and are asymptotes for the root locus, because the pole $s = -1$ scarcely adds to the required phase.

Sec. 6.3] Root locus construction by means of phase lines 113

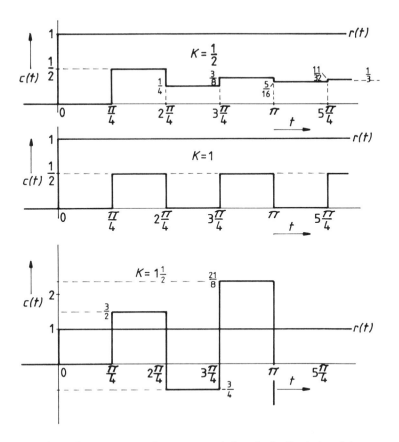

Fig. 6.8 — Step responses of a system consisting of a feedback time delay.

Therefore we may expect that there will be a starting point on the real axis; this we may calculate as follows:

$$\frac{dK'}{ds} = \frac{d}{ds}\{-(s+1)e^s\} = 0 \tag{6.16}$$

Now we have:

$$e^s + (s+1)e^s = 0 ,$$

or:

$$s = -2 .$$

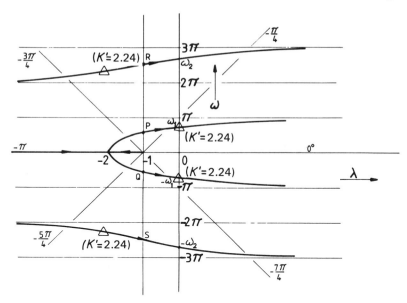

Fig. 6.9 — Root locus of a first order system with time delay.

Thus the two most dominant root locus sections are between the phase lines π and $-\pi$ of the delay. Other points that are also part of the root locus, for various values of K', are, for example, R and S, for here the total phase is also $-\pi$.

For the root locus section passing through point R, the asymptotes is the phase line -2π which is far to the left in the s-plane; while far to the right in the s-plane the asymptote is the -3π phase line of the time delay. A similar consideration applies for the root locus section passing through point S; here the asymptotes are the phase lines 2π and 3π respectively, of the root locus. The remaining root locus sections are always shifted -2π or -3π with regard to the other root locus sections.

Now we may wonder when this system becomes unstable. Instability occurs as soon as there is one pair on the imaginary axis. For such a pole pair:

$$s = j\omega_1 \quad \text{and} \quad s = -j\omega_1.$$

For $s = j\omega_1$ the root locus equation becomes:

$$\frac{e^{-j\omega_1}}{j\omega_1 + 1} = -\frac{1}{K_1} \tag{6.17}$$

From this we obtain two equations:

$$\left| \frac{e^{-j\omega_1}}{j\omega_1 + 1} \right| = \left| -\frac{1}{K_1} \right| \Rightarrow K_1 = \sqrt{1 + \omega_1^2} \tag{6.18}$$

and:

$$\arg\left\{\frac{e^{-j\omega_1}}{j\omega_1+1}\right\} = -\pi \Rightarrow -\omega_1 - \arctan\omega_1 = -\pi \tag{6.19}$$

From (6.19) it follows (by numerical solution), that for example, $\omega_1 \approx 2$ rad/s, and one through this root locus gain at oscillation $K_1 \approx 2.24$ (see also Fig. 6.9). Except for $\omega_1 = 2$ rad/s, more values for ω follow from (6.19), for example $\omega = \pm 8$ rad/s. Substituting this in (6.18) yields the matching value for K: $K_2 = 8$. For $K' = 2.24$ these poles are still in the left half plane at the positions indicated in Fig. 6.9. For greater values of K' than K_1 more pole pairs may cross the imaginary axis. These values of K' may be determined by substituting the matching phase in the phase condition that follows from the root locus equation (see e.g. (6.15)). Usually there will be no interest in these values of K', beacuse then the system is already unstable. The system behaviour is largely determined by the dominant root locus sections between the phase lines $-\pi$ and $+\pi$ of the delay.

6.4 APPROXIMATION METHODS OF A TIME DELAY

From (6.3) it follows that a time delay develops by infinitely many first order systems in cascade. In practice we may also approximate the effect of a delay by a limited number of first order systems in cascade. In expression (6.3) the finite number of first order systems is substituted, hence the following approximations:

$$n = 1: \quad H_d(s) \simeq \frac{1}{1+T_d s} \tag{6.20}$$

$$n = 2: \quad H_d(s) \simeq \frac{1}{\left(1+\frac{T_d}{2}s\right)^2} \tag{6.21}$$

$$n = 3: \quad H_d(s) \simeq \frac{1}{\left(1+\frac{T_d}{3}s\right)^3} \text{ etc.} \tag{6.22}$$

Fig. 6.10 shows an impression of the step responses $u(t)$ belonging to the ideal delay ($n = \infty$) and the approximations.

The phase line method and the approximation of a delay by the first order system do not appear to be very suitable for constructing root loci of systems with delay with the aid of a computer. Another approximation of the delay is the well known Taylor expansion, as in:

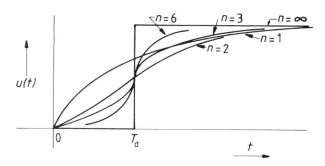

Fig. 6.10 — Approximation of a time delay by means of a cascade of first order systems; step responses.

$$e^{-sT_d} = 1 - sT_d + \frac{s^2 T_d^2}{2!} - \frac{s^3 T_d^3}{3!} + \ldots \qquad (6.23)$$

The position where the sequence is interrupted determines the accuracy of the approximation; the more terms, the more accurate. For small values of the delay T_d the speed of convergence is high; for large values of T_d, respectively many terms of the sequence of (6.23) must be considered.

One of the most widely utilized approximation methods is the Padé approximation. For this purpose the delay is written as the quotient of two polynomials $P(sT_d)$ and $Q(sT_d)$:

$$e^{-sT_d} = \frac{P(sT_d)}{Q(sT_d)} = \frac{a_0 + a_1 sT_d + a_2(sT_d)^2 + a_3(sT_d)^3 + \ldots}{b_0 + b_1 sT_d + b_2(sT_d)^2 + \ldots} \qquad (6.24)$$

The accuracy of this approximation depends on the order of the polynomials P and Q.

The coefficients a_i and b_i may be obtained by dividing the form according to (6.24) and by equalizing it to the expansion. Thus, for a first order approximation it follows that:

$$e^{-sT_d} \simeq \frac{1 - \tfrac{1}{2} sT_d}{1 + \tfrac{1}{2} sT_d} \qquad (6.25)$$

For a second order approximation we have:

$$e^{-sT_d} \simeq \frac{12 - 6sT_d + (sT_d)^2}{12 + 6sT_d + (sT_d)^2} \qquad (6.26)$$

And for a third order system:

$$e^{-sT_d} \simeq \frac{120 - 60sT_d + 12(sT_d)^2 - (sT_d)^3}{120 + 60sT_d + 12(sT_d)^2 + (sT_d)^3} \tag{6.27}$$

etc.

It will be clear that a higher order approximation of a delay will yield a more accurate approximation, but also a more a complex expression, and will thus take more computing time. In Fig. 6.11 we show the computer plot of the root locus of a

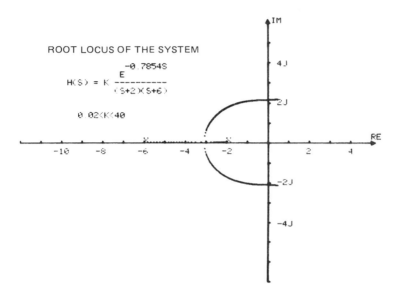

Fig. 6.11 — Root locus (partially) of a second order system with time delay.

system with two poles and a time delay. Only the two dominant root locus sections have been calculated and drawn. For this calculation a second order Padé approximation was used, according to (6.26).

6.5 PROBLEMS

1. Construct, by means of phase lines and the phase criterion of points on the root locus only, the root locus of a system according to the block diagram of Fig. 6.4 in which $H_1(s) = \dfrac{K}{(s\tau + 1)^3}$. Test the result by the construction discussed in Chapter 5.
2. Construct the root locus of the system in Fig. 6.12 for $0 < K < \infty$.
   Also determine the oscillation frequency for this system and the matching value for K.
3. A system, configured as in the block diagram of Fig. 6.4, has in the forward path a

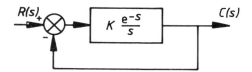

Fig. 6.12 — Block diagram of the system in Problem 2.

gain factor K, a second order system, and a time delay. For the second order system there is a dc gain of 10 and a damping ratio of 1. The undamped frequency is 50 rad/s. The time delay is 0.1 second. Sketch both dominant root locus sections for a variation of K and determine when this system will become unstable and the value of the oscillating frequency.

4. Given the system of Fig. 6.13.

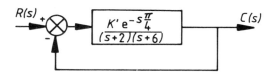

Fig. 6.13 — Feedback second order system with time delay.

Determine K_{osc} and ω_{osc} by estimating ω_{osc}, from the root locus for variation of K' from 0 to ∞.

7
Designing controlled systems in the *s*-domain

7.1 INTRODUCTION

As has been observed in Chapter 1, by means of control engineering we intend to achieve a diversity of objectives. To the control engineer the conversion of these objectives into manageable design criteria means that two system characteristics are essential. These two are the dynamic and the static behaviour of the system.

The block diagram of a controlled system is shown in Fig. 7.1. We have assumed

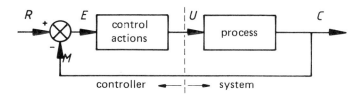

Fig. 7.1 — Block diagram of a controlled process.

the measuring signal fed into the controller to be equal to the output value of the process. If this is not the case, the block diagram may still be remodelled into a block diagram as shown in Fig. 7.1, see Fig. 5.2(a) and (b), but this is not essential. The set value (also set point or reference value) also indicated as a desired value is indicated by R.

The measured value M here equals the controlled value C. In the comparison unit (subtractor) the measured value is compared to the desired value. The difference between these signals is the error signal E. The static value of this signal is indicated by steady state difference or offset.

The control actions are chosen in such a way that from the error signal such a control signal is formed that the controlled value will satisfy the set value so far as possible. For regulator systems we strive to keep the controlled value as constant as possible, in servo systems the output signal of the controlled process should follow the varying reference signal as closely as possible.

7.2 DESIGN CRITERIA IN THE s-DOMAIN

The criteria utilized in designing controlled systems in the s-domain are directly related to properties in the t-domain. These system properties are best determined by a step response, see Fig. 7.2.

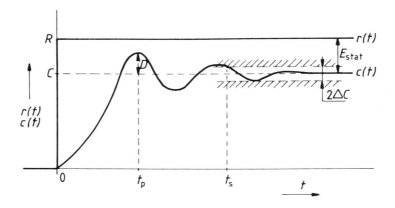

Fig. 7.2 — Design criteria indicated in the step response.

1. The settling time t_s is the time that passes before the step response reaches and stays within a specific band ΔC round the final value of this step response. ΔC is usually expressed by a certain percentage of the final value C.
   We distinguish:

$\Delta C/C = \pm 2\%$: t_s (2%), the indication time.
$\Delta C/C = \pm 5\%$: t_s (5%), the response time.

If a specific value for t_s is required, this may be achieved, as we have already seen in Chapter 4, by imposing a specific requirement on the absolute damping. This requirement follows from:

$$e^{\lambda_2 t_s(2\%)} \leq \frac{1}{50}, \text{ or: } \lambda_2 \leq \frac{-4}{t_s(2\%)} \tag{7.1}$$

and

Sec. 7.2]   **Design criteria in the s-domain**   121

$$e^{\lambda_5 t_s(5\%)} \leqq \frac{1}{20}, \text{ or: } \lambda_5 \leqq \frac{-3}{t_5(5\%)} \tag{7.2}$$

The requirement for the settling time indicates that the two most dominant poles, which are complex conjugate, must not be to the right of the lines of absolute damping $\lambda = \lambda_2$ and $\lambda = \lambda_5$ respectively.

2. The overshoot D is expressed as a percentage of the final value of the step response. For the overshoot of dominant second order systems, we have, according to (4.18):

$$D = e^{(\lambda/\omega)\pi} \cdot 100\% \tag{7.3}$$

If the overshoot must be limited, then a specific requirement is to be imposed on factor λ/ω. This means that the dominant complex conjugate poles must be to the left of the relative damping lines as determined by the requirement. Now the following practical values apply:

$$D \leqq 0.2\%, \text{ then: } \frac{\lambda}{\omega} \leqq -2$$

$$D \leqq 4.3\%, \text{ then: } \frac{\lambda}{\omega} \leqq -1 \tag{7.4}$$

$$D \leqq 11\%, \text{ then: } \frac{\lambda}{\omega} \leqq -0.7$$

$$D \leqq 20.7\%, \text{ then: } \frac{\lambda}{\omega} \leqq -\tfrac{1}{2}$$

The overshoot occurs when $t = t_p$, the peak-time, where:

$$t_p = \frac{\pi}{\omega} \tag{7.5}$$

where ω is the frequency of the damped sinusoidal transient phenomenon.

3. The offset, static error or steady state difference E_{stat} is determined by the value of the dc gain in the open loop of the system, and is expressed as a percentage of the value of the input signal. Thus:

$$E_{stat}(\%) = \frac{100\%}{1 + K_L} \tag{7.6}$$

where K_L is the dc gain in the loop.

If it is required that the offset is zero, then at least one pole of the open loop must be in the origin of the s-plane; for then $K_L = \infty$. If the system to be controlled has no such pole itself, the controller must be provided with an action that produces this pole. A (complete) I-action produces such a pole in zero. Generally, for an offset of zero, the integration must be installed before the point of impact of the disturbance.

Imposing a requirement concerning only the settling time or only the overshoot is no sufficient guarantee that the controlled system is sufficiently damped to make it functional. Usually, both requirements will be imposed simultaneously. This results in declaring part of the s-plane a forbidden area for the two most dominant complex conjugate poles of the system; see Fig. 7.3, in which the relative damping lines for $\beta = \frac{1}{2}\sqrt{2}$ have been drawn, and also the absolute damping line $\lambda = -\lambda_2$.

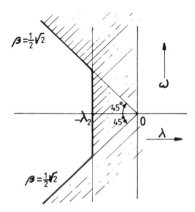

Fig. 7.3. — Forbidden area for the dominant poles.

7.3 DESIGNING A PROPORTIONALLY CONTROLLED SYSTEM

In a proportionally controlled system the relation between the error signal and the output signal of the controller is proportional, or:

$$\frac{u(t)}{e(t)} = K_R \quad \text{and:} \quad \frac{U(s)}{E(s)} = K_R \qquad (7.7)$$

where K_R is the proportional gain.

Usually we assume standard controllers, of which the input range (the range of signal M in Fig. 7.1) and the output range (the range of signal U in Fig. 7.1) are the same. For electronic controllers we usually find ranges of 0 to 10 V, 0 to 20 mA, or 4 to 20 mA. For pneumatic controllers ranges are usually from 0.2 to 1 Pa.

The scale of the controller is, with regard to the adjustment of the proportional

Sec. 7.3]   **Designing a proportionally controlled system**   123

action, usually ganged in a percentage proportional gain. By this proportional gain we understand the part of the input range that must be completed for full control of the output range. Here:

$$PG(\%) = \frac{100\%}{K_R} \qquad (7.8)$$

Example 7.1
A controller has an input and output range of 4 to 20 mA. At a certain setting the measuring signal varies from 10 to 12 mA, and, simultaneously, the output signal of the controller varies from 10 to 4 mA.

Then the proportional factor is 3 and the proportional area is $33\frac{1}{3}\%$; the controller inverts the measuring signal subcontractor. Many controllers are provided with a switch with which the polarity of the system may be reversed. This enables us to achieve feedback in a process in which the output signal decreases when the input signal increases.

Design of P-controlled system is often based on criteria for $t_s(2\%)$ or $t_s(5\%)$ and the overshoot. If the process itself has no integrating effect, there will always be a static error in a proportionally controlled system. The design amounts to taking into account the limitations, following from the design criteria, for the position of the poles of the closed loop system in the s-plane. The procedure will be explained by two examples.

Example 7.2
Given a proportionally controlled system as in the block diagram of Fig. 7.4, the root locus of this system for variation of the proportional gain factor K_R from zero to infinity is sketched in Fig. 7.5.

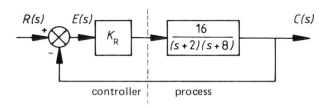

Fig. 7.4 — Block diagram of a proportionally controlled second order system.

The starting point is in -5; the poles of the system are for $K_R = \frac{9}{16}$, both at this starting point. For $K_R > \frac{9}{16}$ the two poles are complex conjugate, and an oscillating transient phenomenon will occur. The absolute damping is constant.
Here, $\lambda = -5$ or $t_s(2\%) \approx 0.8$ second and $t_s(5\%) \approx 0.6$ second.

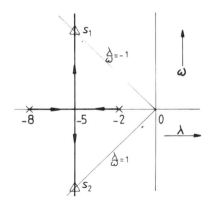

Fig. 7.5 — Root locus of the system of Fig. 7.4.

If the value of K_R increases from the overshoot will increase continually. If we require the overshoot to be limited to 4%, then K_R may increase until the poles are $s_1 = -5 + j5$, $s_2 = -5 - j5$, and $\beta = \frac{1}{2}\sqrt{2}$. The value of K_R is: $\frac{17}{8}$ (P.G. $\approx 47\%$). The offset of the system becomes:

$$E_{\text{stat}}(\%) = \frac{100\%}{1 + \frac{17}{8} \cdot 1} = 32\%$$

Example 7.3
A system as shown in Fig. 7.6 must be set so that the overshoot will be about 20%.

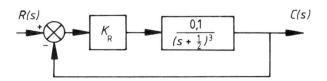

Fig. 7.6 — Block diagram of the system of Example 7.3.

Determine the settling time.
The root locus for variation of K_R is sketched in Fig. 7.7.
The limitation of the overshoot to 20% imposes the requirement for the s-plane, that the root locus may not cross the indicated relative damping lines for $\lambda/\omega = \pm\frac{1}{2}$ ($\beta = 0.45$).
On the $\lambda/\omega = \pm\frac{1}{2}$ line we have $s = -\lambda + 2j\lambda$. Substitution of these values of s in the root locus equation yields:

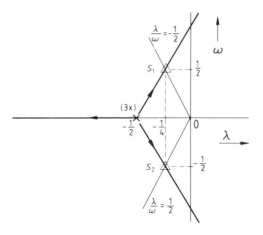

Fig. 7.7 — Root locus of the system of Example 7.3.

$$\frac{0.1}{(-\lambda + 2j\lambda + \frac{1}{2})^3} = -\frac{1}{K_R}$$

from which follows:

$$11 \lambda^3 - 4\tfrac{1}{2} \lambda^2 - \tfrac{3}{4} \lambda + \tfrac{1}{8} + j(-2 \lambda^3 - 6 \lambda^2 + \tfrac{3}{2} \lambda) = -0.1 K_R$$

Now we have:

$$2 \lambda^3 + 6 \lambda^2 - \tfrac{3}{2} \lambda = 0 .$$

$\lambda_1 = 0$ (does not satisfy, belongs to $K_R = 0$)

or

$\lambda^2 + 3 \lambda - \tfrac{3}{4} = 0 \rightarrow \lambda_2 = -\tfrac{3}{2} + \sqrt{3} \simeq 0.23$

$\lambda_3 = -\tfrac{3}{2} - \sqrt{3} \simeq$
$\simeq -3.23$ (does not satisfy, belong to $K_R < 0$)

In our case only $\lambda = 0.23$ satisfies the requirement.

Then $s_1 = -0.23 + j\, 0.46$
and $s_2 = -0.23 - j\, 0.46$.

The third pole s_3 is at $s = -1.04$ (summation rule) with which we obtain from the root locus equation: $K_R = 1.57$ ($PG \simeq 64\%$).

The value of $t_s(5\%)$ is minimal at this value for K_R, as the overshoot borders on the permissible. We find:

$$t_s(5\%) \approx \frac{-3}{-0.23} \approx 13 \text{ sec.}$$

The steady state error is in this case:

$$E_{stat}(\%) = \frac{100\%}{1 + 1.57 \times 0.8} \approx 44\%$$

This example shows that it is impossible to require that in this system $t_s(2\%) < 8$ s or $t_s(5\%) < 6$ s. Please check!

From the above the following conclusions may be drawn for a proportionally controlled system:

1. The values to be achieved for the settling time are strongly dependent on the system parameters.
2. The overshoot may be limited to any value, so a good starting point for designing a proportionally controlled system is the determination of a maximum value for the overshoot.
3. If the process that must be controlled has no pole in the origin by itself, then there will always remain an offset, which means that also in the static situation the output signal, at a step input, will not become quite equal to the reference signal.

7.4 THE ADDITION OF A DERIVATIVE CONTROL ACTION

A derivative control action or lead compensation is applied to increase the damping and/or the response velocity of a system. An increase of the damping – absolute and/or relative – is achieved, as we can see in the frequency domain, by the phase lead produced by such action.

The increase of the response velocity will then be at the expense of the extra amplification for higher frequencies, in this case the increased bandwidth, achieved by the derivative action.

The modulus and phase characteristics Bode diagram of a derivative action are shown in Fig. 7.8, from which appear the phase lead (in a specific frequency area) and the high frequency gain. The transfer function is:

$$H_d(j\omega) = \frac{1 + j\omega\tau_d}{1 + j\omega\frac{\tau_d}{a}}, \tag{7.9}$$

here τ_d is the differential time and a the taming factor. In practice we use this 'tamed' derivative action to limit the extra high frequency gain. Thus undesired disturbances of a higher frequency are not amplified too much. Moreover, the phase lead is

Sec. 7.4] The addition of a derivative control action

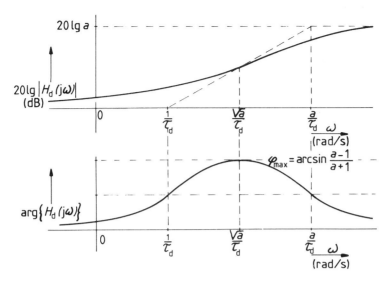

Fig. 7.8 — Bode diagram of a derivative action.

only necessary in a specific frequency area. Usually we choose values 6, 10, or 20 for the taming factor.

The transfer function may be derived from (7.9), and becomes:

$$H_d(s) = a \frac{s + \dfrac{1}{\tau_d}}{s + \dfrac{a}{\tau_d}} . \quad (7.10)$$

The pole/zero plot is shown in Fig. 7.9.

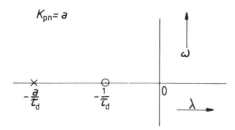

Fig. 7.9 — Pole/zero plot of a derivative action.

We shall now explore the influence of a differentiating action on the procedure whereby the value of τ_d is varied in such a way that the zero-pole combinations moves from left to right along the real axis. Let us assume a system as shown in Fig. 7.10.

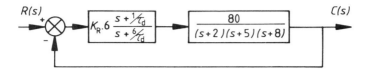

Fig. 7.10 — Block diagram of a PD controlled system.

For $\tau_d = 0$ the zero-pole combination is at infinity; the derivative action does not influence the system behaviour. The root locus for variation of K_R then looks like the one in Fig. 7.11a.

For $\tau_d = \frac{1}{12}$ s only slight influence occurs, see Fig. 7.11b.

For $\tau_d = \frac{1}{8}$ s the origin of the D action exactly compensates for the system pole at $s = -8$; then also the root locus of the controlled system will be scarecely different (see Fig. 7.11c) from the previous two cases.

Because we choose for $\tau_d = \frac{1}{5}$ second the system pole at $s = -5$ is compensated. The root locus has now been shifted to the left, so that the settling time decreases. An even further improvement occurs if the system pole at $s = -2$ is compensated by $\tau_d = \frac{1}{2}$ second, (see Fig. 7.11d and e).

Of course it is theoretically possible to compensate a pole exactly by choosing τ_d through the zero of the differentiating action. In practice we must assume this to be difficult. Even if we could succeed in compensating a pole (its position must be exactly known), we still have to take into account that the position of such a pole will vary because of, for example, the influence of temperature on specific system parameters. Thus complete compensation is overridden.

In Fig. 7.11f, g, and h three cases are shown in which the zero is next to the pole, to the right of the pole at $s = -5$. The characteristic of these cases is that there will be an extra root locus section on the real axis to the right of the pole at $s = -5$. For the time response this means that an e-function will develop, determined by the pole of the controlled system on this root locus section. Because of the position of this pole this implies a component in the time response that phases out relatively slowly.

Finally, in Fig. 7.11i we find the case in which the zero of the differentiating action has been intentionally put slightly to the left of the pole at $s = -5$.

Now the root locus is situated conveniently so far as the damping is concerned, because the root locus sections are as it were pulled a little into the left half-plane by the zero, while the extra pole of the closed loop system is further to the left, viz. on the root locus section between $s = -6$ and $s = -8$.

A rule of thumb for choosing τ_d is that for systems with two or more poles the differentiating time is chosen to be a little less than the second largest time constant of the system.

Sec. 7.4] The addition of a derivative control action

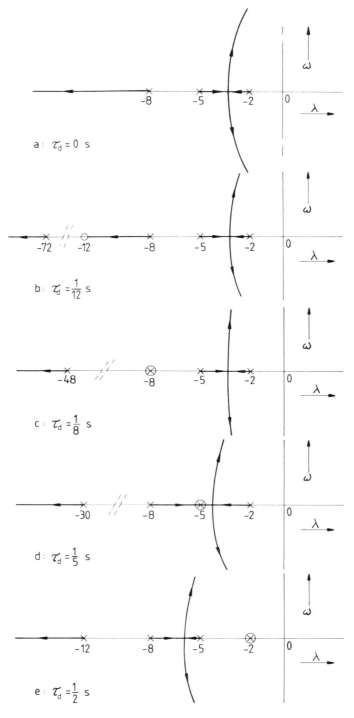

Fig. 7.11 — Root loci for different values of τ_d (continued next page).

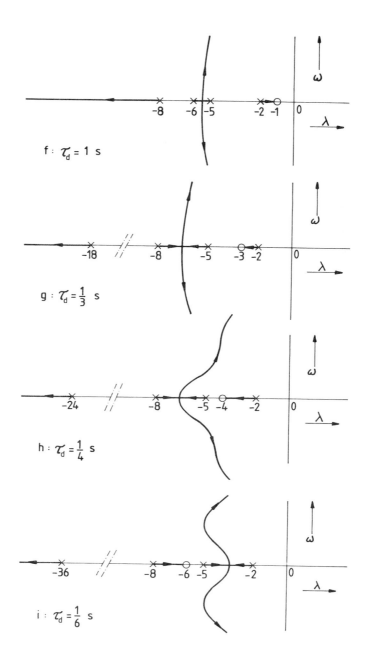

Fig. 7.11 (continued) — Root loci for different values of τ_d.

The degrees in which τ_d is chosen to be less, depends on the variation to be expected in the location of the pole belonging to the second largest system time constant.

In sections 7.7 and 7.8 a more accurate method for designing a derivative control action will be introduced.

7.5 THE ADDITION OF AN INTEGRAL CONTROL ACTION

As has been mentioned before, a P-controlled system will always retain an offset if at least the system itself has no pole at the origin. However, for zeroing the offset it is necessary that the static loop gain becomes infinitely large.

Assuming a system without a pole at the origin we may now add an integrating control action to the controller, to satisfy our objective of zeroing the offset.

In the ω-domain the transfer function of such an I-action is

$$H_i(j\omega) = 1 + \frac{1}{j\omega\tau_i} = \frac{1 + j\omega\tau_i}{j\omega\tau_i}, \text{ in which } \tau_i \text{ is the integration time.} \quad (7.11)$$

The Bode diagram is shown in Fig. 7.12. The dc gain is indeed infinite, as the

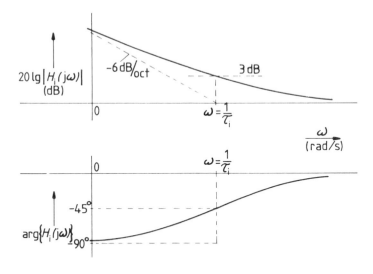

Fig. 7.12 — Bode diagram of an integral control action.

high frequency gain is one. This actions produces an (extra) negative phase contribution in the lower frequencies, which usually presents no danger to the stability of the system.

The transfer function becomes:

$$H(s) = \frac{s + \dfrac{1}{\tau_i}}{s} \qquad (7.12)$$

and the pole/zero representation is as shown in Fig. 7.13.

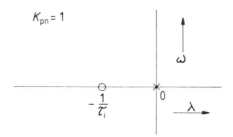

Fig. 7.13 — Pole/zero plot of an integral control action.

The choice of the integrating time τ_i will be illustrated by the system of Fig. 7.14.

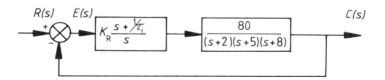

Fig. 7.14 — Block diagram of a PI controlled system.

If we choose $\tau_i = \infty$, then the pole and the zero of the integral action will neutralize each other, and the system will be proportionally controlled. The root locus for variation of K_R will be as shown in Fig. 7.15a. In Fig. 7.15b we have set τ_i at one second; the zero pulls the dominant root locus sections a little toward the imaginary axis. The pole of the closed loop system, between $s = 0$ and $s = 1$, causes a relatively slow transient phenomenon in the response.

If we choose $\tau_i = \frac{1}{2}$ second, see Fig. 7.15c, the zero at $s = -2$ will compensate for the pole being there too. At the first this assumption seems convenient; in practice, however, we must again take into account the variations of the location of the poles. For values of τ_i that are smaller than $\frac{1}{2}$ second the starting point of the root locus sections on the real axis will move to the right considerably, see Fig. 7.15d, in which $\tau_i = \frac{1}{3}$ second). Although the root locus sections, as soon as they have left the real axis, are attracted by the zero, a considerably worse damped system will develop.

Summarizing, we may say that in all the cases shown in Fig. 7.15 (except for (a))

Sec. 7.5] The addition of an integral control action

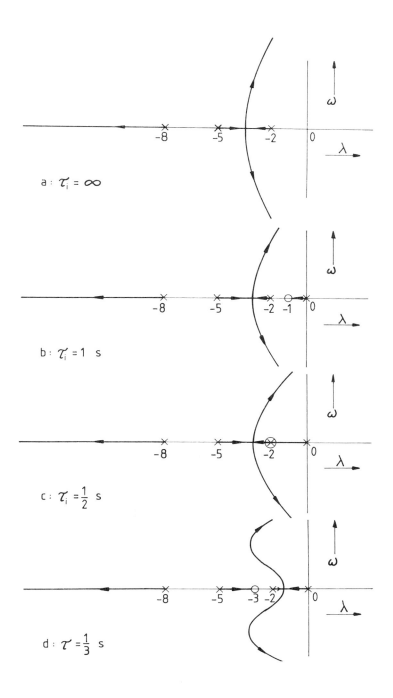

Fig. 7.15 — Root loci of a PI controlled system.

the offset will be zeroed completely. The larger we choose τ_i to be, the longer the integration of the offset will take.

Consequently, the right choice for τ_i is the one equal to or a little larger than the largest time constant of the system.

7.6 ROOT LOCI IN FLUCTUATION OF A TIME CONSTANT

Up till now we have always considered the path of the root locus in fluctuation (variation) of a gain factor. However, it is also possible to determine the path of the root locus when another parameter in the control loop varies. We shall illustrate this by means of three cases, in which we shall add to the system a zero, a pole, and a pole/zero combination respectively. The additional parameter will be varied, and the path of the root locus will be examined.

Of course it is also possible, by means of a root locus, to determine to what extent a system is affected by changes in systems parameters, such as wear, temperatures, etc.

1. Adding and varying a zero

In Fig. 7.16 we have a system to which a zero $s = z_i$ is added in such a way that the DC

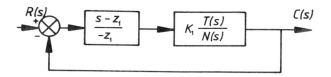

Fig. 7.16 — System with additional zero.

gain in the loop does not change as a result of this addition.
The root locus becomes:

$$1 + \frac{s - z_1}{-z_1} \cdot K_1 \frac{T(s)}{N(s)} = 0 \qquad (7.13)$$

and also:

$$\frac{N(s) + K_1 T(s)}{s K_1 T(s)} = -\frac{1}{-z_1} \qquad (7.14)$$

Naturally, the same rules apply for this root locus as those we derived and utilized for root loci for which K_1 was variable. The starting points of this root locus we find to be for $z_1 = -\infty$; the zero will then be at $-\infty$, and does not affect the poles of this system. We obtain these starting points from (7.14) by assuming:

Sec. 7.6] Root loci in fluctuation of a time constant

$$N(s) + K_1 T(s) = 0 \tag{7.15}$$

This is the root locus equation for K_1, without additions. So the starting points are on the 'normal' root locus (for variation of K_1), in positions belonging to the value for $K = K_1$. The arrival points of the roots locus for variation of z_1 are to be found for $z_1 = 0$; these arrival points are obtained from (7.14):

$$s \; K_1 \; T(s) = 0 \tag{7.16}$$

So for $z_1 = 0$, the root locus sections of the root locus for a variation of z_1 from $-\infty$ to zero, will end in the possible zeros of the original system, while there is one (extra) zero in the origin.

Example 7.4
Let us assume the system of Fig. 7.17; in this system the additional zero z_1 is varied

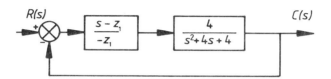

Fig. 7.17 — The system according to Example 7.4.

from $-\infty$ to 0.

The root locus equation will be:

$$1 + \frac{s - z_1}{- z_1} \cdot \frac{4}{s^2 + 4s + 4} = 0$$

and also:

$$\frac{s^2 + 4s + 8}{4s} = -\frac{1}{-z_1}$$

The starting points for $z_1 = -\infty$ are at $s_1 = -2 + j2$ and $s_2 = -2 - j2$ respectively, as we find one arrival point for $z_1 = 0$ at $s_3 = 0$; the second point of arrival is assumed to be at infinity.

In Fig. 7.18 the zero at the origin of this root locus is indicated by a small square, and the starting points by small triangles; the latter being the poles of the closed loop system, without additions.

One of the two poles of the original system will proceed to the arrival in the

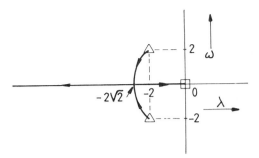

Fig. 7.18 — Root locus for varying the additional zero from minus infinity to zero.

origin, and the other one will proceed to $-\infty$. According to the construction rules for root loci the part of the real axis to the left of the origin belongs to the root locus. There must be a point of arrival on the real axis, for which we find: $s = -2\sqrt{2}$. Please check!

It is clear that the root locus that has developed, yields valuable information. Note that by increasing the zero from $-\infty$ to 0, the damping of the system will increase from the under-critical ($\beta < 1$) through critical $\beta = 1$ arrival point to the over-critical damping ($\beta > 1$). Therefore we must note that just a zero (for this is a pure differentiation) may not be realized in practice, so that often a zero-pole combination (lead compensation) is applied to increase the damping.

2. *Adding and varying a pole*

Our starting point is the system of Fig. 7.19 to which a pole has been added in such a

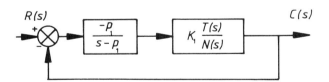

Fig. 7.19 — Block diagram of a system with additional pole.

way that again the dc gain has not changed.

For the root locus equation we may derive:

$$\frac{N(s) + K_1 T(s)}{sN(s)} = -\frac{1}{-p_1} \quad \text{(Please check!)} \tag{7.17}$$

The starting points of the root locus for variation of $-p_1$ (the extra pole) are found for $p_1 = -\infty$ at the poles of the closed loop system. The arrival points are found for $p_1 = 0$ at the poles of the open loop system, with an extra arrival at $s = 0$.

Example 7.5
Given the system of Fig. 7.20; the additional pole is varied from $-\infty$ to 0.

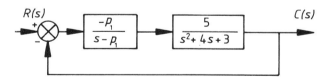

Fig. 7.20 — The system of Example 7.5.

The root locus equation for variation of p_1 will be:

$$\frac{s^2 + 4s + 8}{s(s^2 + 4s + 3)} = -\frac{1}{-p_1}$$

Starting points ($p_1 = -\infty$): $s_1 = -2 + j2$ and $s_2 = -2 - j2$.
Arrival points ($p_1 = 0$) are: $s_3 = 0$, $s_4 = -1$, and $s_5 = -3$.
The root locus is shown in Fig. 7.21.

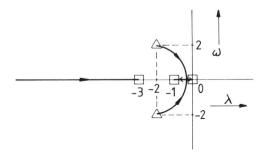

Fig. 7.21 — Root locus for variation of the additional pole from $-\infty$ to 0.

The conclusion must be that when the extra pole comes closer to the other poles at $s = -1$ and $s = -3$ the system will be less damped, as could be expected.

3. Adding a pole/zero combination
We shall consider only the case in which zero is to the right of the pole and that there is a set relation ($a > 1$) between the positions of that pole and the zero: the lead compensation or derivative action has been mentioned before (see section 7.4). The block diagram is shown in Fig. 7.22.

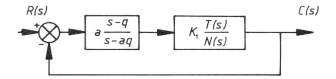

Fig. 7.22 — Block diagram of a system with additional pole/zero combination.

From the root locus equation:

$$1 + a \frac{s-q}{s-aq} \cdot K_1 \cdot \frac{T(s)}{N(s)} = 0 \qquad (7.18)$$

follows the root locus equation in normal from:

$$\frac{a\{N(s) + K_1 T(s)\}}{s\{N(s) + a K_1 T(s)\}} = -\frac{1}{-q} \qquad (7.19)$$

For $q = -\infty$ the extra pole/zero combination is at $-\infty$; the starting points of the root locus are found by assuming:

$$N(s) + K_1 T(s) = 0. \qquad (7.20)$$

Note that this is the root locus equation of the system without additional pole/zero combination. So the starting points are the poles of the original closed loop system for specific value for K_1. We obtain the arrival points for $q = 0$ from:

$$s\{N(s) + a K_1 T(s)\} = 0 \qquad (7.21)$$

One arrival is at $s = 0$; the others are on the root locus of the original system, matching a root locus gain $K' = aK_1$. Note that the starting points as well as the arrival points, except the one at $s = 0$, are on the root locus of the original system. The arrival points in question are further down the root locus path.

Example 7.6
Let us assume the system of Fig. 7.23; in Fig. 7.23a we have the block diagram of a PD controlled first order process in the ω-domain, while Fig. 7.23b shows the same process in the s-domain.

Note that in this case, compared to (7.18), we have $q = -\frac{1}{\tau_d}$ and $a = 10$. The root locus equation becomes:

$$1 + 10 K \cdot \frac{s + \frac{1}{\tau_d}}{s + \frac{10}{\tau_d}} \cdot \frac{20}{s + 10} = 0 \qquad (7.22)$$

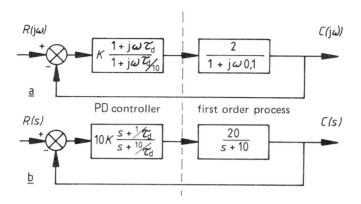

Fig. 7.23 — Block diagram of a PD controlled first order process.

For the root locus equation with variable τ_d we obtain from (7.22):

$$\frac{s(s+10+200K)}{10(s+10+20K)} = -\frac{1}{\tau_d} \tag{7.23}$$

Now assume that the system without D action $\tau_d = 0$ is proportionally controlled in such a way that $K = 1$, then:

$$\frac{s(s+210)}{10(s+30)} = -\frac{1}{\tau_d} \tag{7.24}$$

For $\tau_d = 0$ the pole/zero combination is at $-\infty$ and does not affect the (proportionally) controlled process. So a starting point of the root locus for variation of τ_d from zero to infinity is at $s = -3$); this is the pole of the P controlled process for $K = 1$. For τ_d we obtain from:

$$s(s+210) = 0 \tag{7.25}$$

two arrival points, viz. at $s = 0$ and $s = -210$.

With the starting point in $s = -30$ and the arrival points at $s = 0$ and $s = -210$ the root locus for variation of τ_d from 0 to infinity, will look, according to the current construction rules, as shown in Fig. 7.24. There are two root locus sections, for there are two poles, viz. one of the process and one of the D action; the second starting point is at $-\infty$.

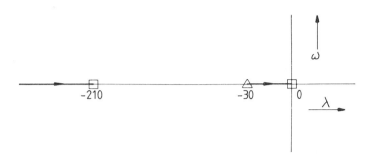

Fig. 7.24 — Root locus for variation of τ_d from 0 to ∞.

When first considering the root locus of Fig. 7.24, we might conclude that the system becomes slower by adding the D action, because the dominant pole will move towards the imaginary axis when τ_d is increased, for the exponent of the e-function in the response because of this dominant pole becomes larger (less negative). However, we must consider that the zero of the D action affects the size of the contribution of the separate poles in the total response.

If we assume, for example, $\tau_d = 0.05$ second and $K = 1$, then the pole/zero representation of the controlled system will be as indicated in Fig. 7.25a; in this case the zero is at $s_1 = -20$ and there are two poles at $s_2 \approx -15$ and $s_3 \approx -395$ respectively, as may be determined from equation (7.24). The impulse response may now be written as:

$$\left.\begin{aligned} c(t) &= A_1 \, e^{-15t} + A_2 \, e^{-395t} \\ \text{with} \quad A_1 &= \frac{+5}{380} = \frac{1}{76} \\ A_2 &= \frac{-375}{-380} \approx 1 \end{aligned}\right\} \quad (7.26)$$

In this case the response is almost completely determined by the pole at $s = -395$, and the system has become substantially faster because of the D action.

If we assume $\tau_d = 0.1$ s, then there are two poles at $s_2 = -10$ and $S_2 = -300$, while the zero at $s_1 = -10$ compensates the pole at s_2, see Fig. 7.25b. Then the system has only one pole at $s_3 = -300$: the system has again become faster because of the D action. Also for larger values τ_d the influence of the pole that is furthest to the right, remains slight in the total response, as the reader can see for himself.

Note
The conclusion from the above analysis, as to the effect of a D action applied in controlling a first order process, must be that this D action speeds up the controlled process. That the application of a D action in controlling a first order process is hardly

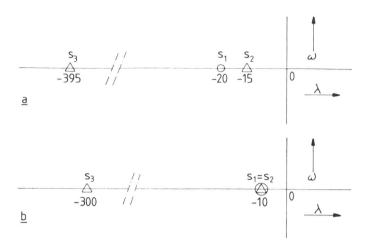

Fig. 7.25 — Pole/zero plots of a PD controlled first order system.

significant after all, lies in the fact that a similar result may be achieved by simply increasing the proportional gain of the P controller. Theoretically the response velocity may then be chosen as large as we might want it to be, while in addition a larger proportional gain will decrease the offset.

7.7 COMPUTER AIDED DESIGN OF A DERIVATIVE CONTROL ACTION

A difficulty that occurs in analysing the effect of τ_d variations on the position of poles in a closed loop system lies in our not being able to easily determine the course of certain root locus sections. In such cases the use of a computer is essential, as appears from the following example.

Example 7.7
Assume the controlled system of Fig. 7.26; the proportional gain of K and the

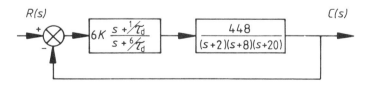

Fig. 7.26 — Controlled system of Example 7.7.

differentiating time τ_d must be set. The settling time must be as short as possible, and the overshoot must be limited to about 10%.

Now, by means of the P action, we shall first set the system such that the overshoot becomes about 10%; next, we shall try to decrease the settling time by the D action.

The root locus equation for varying K by τ_d may be written as:

$$\frac{448}{(s+2)(s+8)(s+20)} = -\frac{1}{K} \qquad (7.27)$$

The root locus is shown in Fig. 7.27.

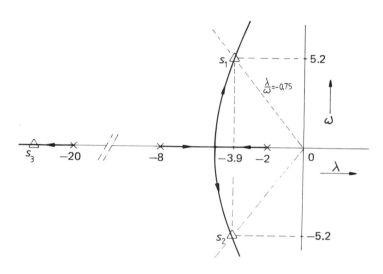

Fig. 7.27 — Root locus of a P controlled third order process.

For $K=1.4$ the system has three poles $s_1 = -22.2$, $S_2 = -3.9 + j5.2$, and $S_3 = -3.9 - j5.2$. The relative damping for the two dominant conjugate poles s_1 and s_2 will become: $\lambda/\omega = 3.9/5.2 \approx 0.75$. For this value of the relative damping the overshoot is about 9.5%. The offset is, for this setting of a P controlled system, about 33% of the input step amplitude. The settling time (2%) is now about 1 sec.

Starting from this setting we need to find, for the proportional control action that we are going to investigate, the effect of the D action (with taming factor 6) on the properties of the process. Therefore the τ_d locus is constructed for variation of τ_d from zerto to infinity. The root locus equation is written as:

$$\frac{s\{(s+2)(s+8)(s+10) + 1.4 \times 6 \times 448\}}{6\{(s+2)(s+8)(s+20) + 1.4 \times 448\}} = -\frac{1}{\tau_d} \qquad (7.28)$$

The starting points of the τ_d locus will be found for $\tau_d = 0$ at the poles of the P

Sec. 7.7] Computer aided design of a derivative control action

controlled process with $K = 1.4$; s_1, s_2, and s_3, and the arrivals for $\tau_d = \infty$ follow from:

$$s\{(s+2)(s+8)(s+20) + 1.4 \times 6 \times 448\} = 0 \qquad (7.29)$$

There is one arrival point at $s = 0$, and furthermore there are three arrivals that result from:

$$(s+2)(s+8)(s+20) + 1.4 \times 6 \times 448 = 0 \qquad (7.30)$$

These arrival points are on the root locus of the system for variation of K, and at those positions on the root locus for which the root locus gain has a value for $K' = 1.4 \times 6$.

For these arrivals we find: $s = 0$, $s = -27.6$, $s = -1.2 + j12$, and $s = -1.2 - j12$. By computer calculation of the poles for various values of τ_d, and by plotting these values, the root locus of Fig. 7.28 will develop.

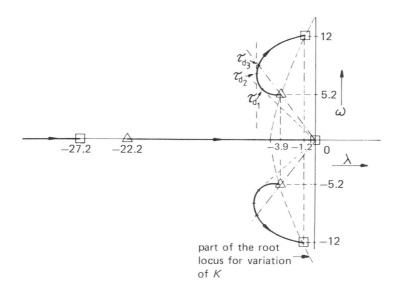

Fig. 7.28 — τ_d locus for the PD controlled process.

From the τ_d locus it appears that there are several favourable settings for τ_d namely:

- $\tau_d = \tau_{d1}$; for this setting, at the line of maximum relative damping, the overshoot will be minimal ($\tau_d = 0.25$ s).
- $\tau_d = \tau_{d2}$: here the overshoot is smaller than required (10%) when the maximum absolute damping is achieved. The settling time (2%) is about half the value of the

P controllled system ($\tau_{d2} \approx 0.3$ s). For τ_{d3} the overshoot of the PD controlled system is equal to the overshoot of the P controlled system; however, the absolute damping is considerably larger.

Notes:
1. With the above determined values for the D action we have not yet achieved optimum setting, for here the choice of the proportional setting is very arbitrary.
   From another setting for K it would be possible to obtain an even more favourable τ_d locus. Also, the choice of the taming factor of the D action effects the τ_d locus. For an optimum setting of the controller, it will be necessary, depending on the objectives for the dynamic behaviour of the system — overshoot and setting time — to solve an optimizing problem with three variables, viz. K, the taming factor, and τ_d.
2. In the same way the D action may be set by means of the τ_d locus; it is also possible to set the integral action by means of the information obtained from the τ_i locus.

7.8 DESIGNING A DERIVATIVE CONTROL ACTION BY MEANS OF THE PHASE CONDITION OF THE ROOT LOCUS

For each point p_j of the root locus we have the phase condition:

$$\sum_{i=1}^{m} \arg(p_j - z_i) - \sum_{i=1}^{n} \arg(p_j - p_i) = \pi \pm 2k\pi \text{ with: } k = 0,1,2,\ldots \quad (7.31)$$

If a certain point in the *s*-plane is not part of the root locus, we may try to make it part of the root locus by adding one or more poles and/or zeros. By this addition the point in question must satisfy the phase condition of (7.31).

Example 7.8
Let us assume the system of Fig. 7.29; first the second order system is set, by means of

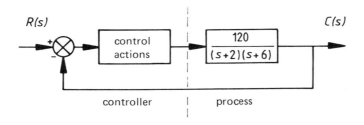

Fig. 7.29 — The controlled system of Example 7.8.

the P action, in such a way that the overshoot will be limited to 4%. Next we design such a D action that the setting time will be halved.
   For the proportionally controlled system the root locus equation for variation of the gain factor K becomes:

Designing a derivative control action

$$\frac{120}{(s+2)(s+6)} = -\frac{1}{K} \tag{7.32}$$

The root locus is sketched in Fig. 7.30. For the given restriction for the overshoot,

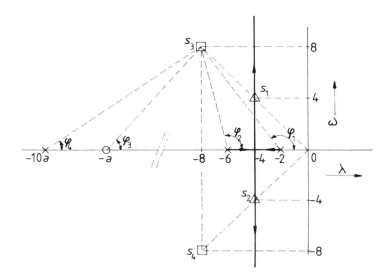

Fig. 7.30 — Root locus of the controlled system of Example 7.8.

K should be limited to $K = \frac{1}{6}$. The poles of the closed-loop system will be at: $s_1 = -4 + j4$ and $s_2 = -4 - j4$. The settling time (2%) is about 1 s ($\lambda/\omega = -1$).

For halving the settling time at the same overshoot, the dominant poles of the system must be positioned at s_3 and s_4. The two dominant root locus sections must pass through these points. Because of the poles at $s = -2$ and $s = -6$ the phase at s_3 is now equal to:

$$\varphi = -(\varphi_1 + \varphi_2) = -231° < -180°. \tag{7.33}$$

Then the pole/zero combination of the derivative action must yield a phase lead at s_3 of:

$$\varphi_d = 180° - \varphi = 51° \tag{7.34}$$

If we take a pole/zero combination as a derivative control action as in

$$H_d(s) = 10 \cdot \frac{s+a}{s+10\,a}, \tag{7.35}$$

then the following must apply (see Fig. 7.30):

$$\varphi_d = 180° - \varphi = \varphi_3\varphi_4 = 51° \tag{7.36}$$

Now we have the phase contribution φ_d of the D action at s_3:

$$\varphi_d = \arctan \frac{8}{a-8} - \arctan \frac{8}{10a-8} \tag{7.37}$$

From the equation:

$$\varphi_3 - \varphi_4 = \arctan \frac{8}{a-8} - \arctan \frac{8}{10a-8} = 51° \tag{7.38}$$

We may determine a. Let us consider:

$$\tan(\varphi_3 - \varphi_4) = \frac{\tan \varphi_3 - \tan \varphi_4}{1 + \tan \varphi_3 \cdot \tan \varphi_4}, \tag{7.39}$$

Then follows:

$$\frac{\dfrac{8}{a-8} - \dfrac{8}{10a-8}}{1 + \dfrac{8}{a-8} \times \dfrac{8}{10a-8}} = \tan 51° \simeq 1.23$$

The solution of this results in the quadratic equation:

$$a^2 - 14.65\, a + 12.8 = 0 \tag{7.40}$$

This equation has two roots, $a_1 \approx 0.9$ and $a_2 \approx 13.7$, of which only a_2 satisfies our problem; from choosing a_1 a totally different root locus will develop, yielding a slower transient phenomenon (see section 7.4, for too large value of τ_d). As $a = 1/\tau_d$, the differentiating time will be set at $\tau_d = \dfrac{1}{a} \approx 0.07$ s.

The amplification of the P action, in the case that the root locus will pass through s_3, may be calculated by substituting s_3 in the root locus equation:

$$\frac{120 \times 10\,(s + 13.7)}{(s + 137)(s + 2)(s + 6)} = -\frac{1}{K} \tag{7.41}$$

Sec. 7.8] **Designing a derivative control action** 147

From this we obtain $K \approx 0.9$.

An extremely good (graphical) method to determine the position of the pole/zero combination of a D action, is shown in Fig. 7.31. First we draw, starting from P where

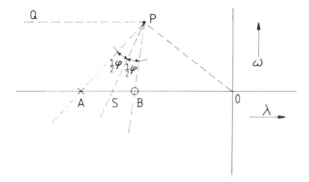

Fig. 7.31 — Graphic determination of a D action.

a certain phase lead φ is desired, the horizontal line PQ, and next the connecting line PO from P to the origin. PS is the bisector of the angle OPQ. At both sides of this bisector we set equal angles of $\frac{1}{2}$ φ. The intersections of PB and PA with the real axis indicate the positions of the zero (B) and the pole (A) of the D action.

This method is based on a maximum ratio b/a. Consequently the amplification of high frequencies (noise) will be minimal. The starting point for the proof of the procedure is shown in Fig. 7.32.

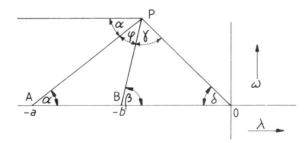

Fig. 7.32 — Starting point for the proof of the method of achieving the optimum D action.

Applying the sinus rule in Fig. 7.32 yields:

$$\frac{PB}{\sin \delta} = \frac{b}{\sin \gamma} \tag{7.42}$$

and

$$\frac{PB}{\sin \alpha} = \frac{a-b}{\sin \gamma} \tag{7.43}$$

From (7.42) and (7.43) follows, with $a = 180° - (\varphi + \gamma + \delta)$ and $\sin x = \sin(180° - x)$:

$$\frac{a-b}{b} = \frac{\sin \varphi \sin \delta}{\sin \gamma \sin(\varphi + \gamma + \delta)}$$

and so:
$$\frac{a}{b} = 1 + \frac{\sin \varphi \sin \delta}{\sin \gamma \sin(\varphi + \gamma + \delta)} \tag{7.44}$$

The ratio b/a is maximal, if the ration a/b is minimal. From (7.44) it follows that is the case if the term $\sin \gamma \sin(\varphi + \gamma + \delta)$ is maximal. Here,

$$2 \sin \gamma \sin(\varphi + \gamma + \delta) = \cos(\varphi + \delta) - \cos(\varphi + 2\gamma + \delta)$$

This is maximal if $\cos(\varphi + 2\gamma + \delta) = -1$, so that:

$$\varphi + 2\gamma + \delta = 180° \tag{7.45}$$

Also: $\quad \varphi + \gamma + \delta + \alpha = 180° \tag{7.46}$

From this follows: $\quad \gamma = \alpha \tag{7.47}$

The necessity to be equal for the angles α and γ leads to the construction shown in Fig. 7.31.

Starting from our example 7.8 a phase lead of 51° must develop at point s_3, because of the pole/zero combination.

The required design procedure is shown in Fig. 7.33. We find the zero at $s = -7.6$

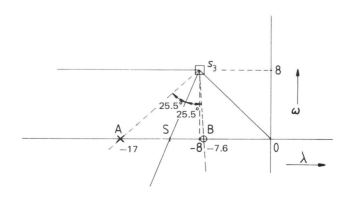

Fig.7.33 — Graphical design of a pole/zero combination.

Sec. 7.8]   Designing a derivative control action   149

and the pole at $s = -17$. Consequently the transfer function of the compensation network becomes:

$$H_d(s) = 2.24 \frac{s+7.6}{s+17}, \text{ the taming factor is 2.24.}$$

The new value for K may again be determined from the complete root locus equation; we obtain $K \approx 0.47$.

In Fig. 7.34a the step responses of the designed system are indicated for the calculated values of the modifying factor a. The control signals belonging to it (output of the controller) are shown in Fig. 7.34b.

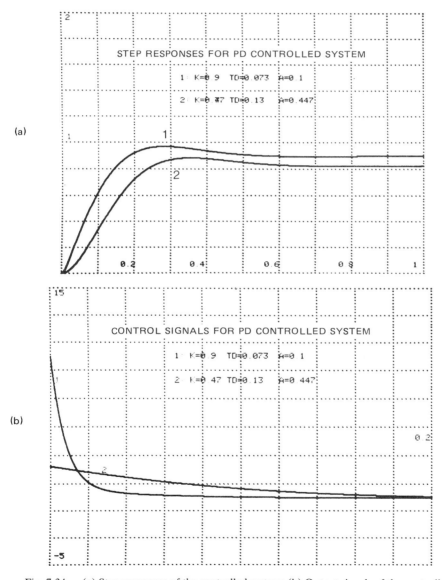

Fig. 7.34 — (a) Step responses of the controlled system: (b) Output signals of the controller.

7.9 PROBLEMS

1. A system consists of 3 identical first order systems in cascade. Each first order system has a time constant of 0.5 s and a dc gain of 2.

   This system is included in a control loop with unit feedback. The controller has only P action with a proportionality factor equal to 0.5.

   Determine for the controlled system $t_s(2\%)$, $t_s(5\%)$, $E_{stat}(\%)$, and the overshoot in a step response.

2. For a system with two dominant complex conjugate poles it is required that $t_s(5\%) \leq 0.5$ s and $D \leq 16\%$. Indicate by shading the forbidden area for these dominant poles.

3. We have the system of Fig. 7.35.

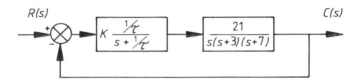

Fig. 7.35 — Block diagram of the system of exercise 3.

Question:
a   Assume $\tau = 0$; determine K so that $t_s(2\%) = 4$ sec.
b   Determine the τ locus for variation of τ from zero to infinite by means of the value for K and its pole positions found in (a).
c   About what value may τ have in order that $t_s(2\%) \leq 8$ sec.

4. A process contains one pure integrator and a first order system with a time constant of $\frac{1}{2}$ sec. and a dc gain of 2. This process is controlled by a controller with one unit feedback of PD control actions.

   Question:
   a   The relative damping factor β must have a value of 0.5. Determine the matching value for the proportionality factor if the process is only proportionally controlled. Also determine $t_s(2\%)$.
   b   It is desired that the value of t_s, found in (a), is halved at the same overshoot by applying a D action. The taming factor is 10. Determine the location of the pole/zero combination by the numerical method described in section 7.8. Also determine the new value for the proportionality factor K.
   c   This is the same question as in (b); now, however, the location of the pole/zero combination is determined graphically. Also determine the new value for K.

Problems

5. We have the control system of Fig. 7.36. It is required that the overshoot in the step response is about 20%, and $t_s(2\%)$ must be about 2 sec.

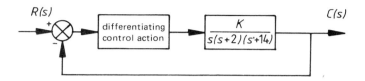

Fig. 7.36 — PD controlled third order system.

a   Determine the location of the dominant poles of the controlled system so as to satisfy the requirements.

b   Prove, by means of a root locus construction, that a proportional control (no D action) will not satisfy the desired requirements.

c   Design a derivative control action with taming factor 10 and transfer function

$$G(s) = 10 \frac{s+a}{s+10\,a}$$

in order to satisfy the requirements.
      Also determine the matching value for K.

d   Design a derivative control action by the graphical method of section 7.8. Utilize the transfer function

$$G(s) = \frac{b}{a} \frac{s+a}{s+b}.$$

Also determine the matching value for K.

8

Three case studies of analog control systems

8.1 INTRODUCTION

In this chapter we shall discuss in detail the control design of a DC position servo system, a phase lock loop, and a 'gravity raiser'.

The DC position servo system and the phase lock loop are examples of designs in which the application of the pole/zero theory is simple. The third example shows us, that especially by utilizing a computer for the application of root loci, a very good setting of control actions may be achieved.

8.2 A DC POSITION SERVO SYSTEM

The objective of a servo system is to control a motor, by means of a (small) control signal in such a way that an object driven by the motor will be moved to a certain position. We might imagine directing an aerial, positioning the pen of a recorder or aiming a gun.

Fig. 8.1(a) is a sketch of a DC position servo system. The servo-motor M, the gearbox, the mechanical load, the tachogenerator T, and the potentiometer R_2 are coupled mechanically. There is a slow axle, on which the potentiometer R_2 has been mounted, and a fast axle to which the tachometer T has been fitted

Potentiometer R_1 gives the desired position θ of the slow axle; potentiometer R_2 gives the position of this axle.

Amplifier V_1 determines the difference between the voltages U_1 and U_2; so voltage U_3 is a measure of the deviation from the desired position and the actual position of the axle. In amplifier V_2 the voltage U_4 is — if necessary — subtracted from U_3; the former is the tachofeedback signal. The output voltage of amplifier V_2 controls the servo balance amplifier, which in its turn feeds both fields of a split field motor. The advantage of controlling a split field motor by two bifilar coils is that remanent magnetism is reduced, so the remaining torque, if $U_5 = 0$, may be very small indeed. The motor armature is fed from a current source.

A DC position servo system

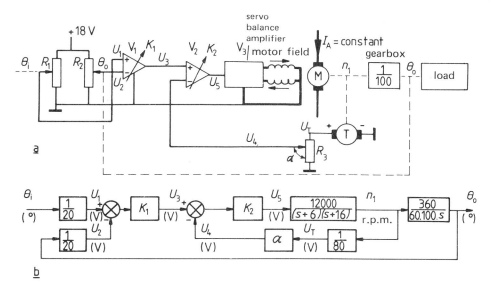

Fig. 8.1 — Sketch and block diagram of a DC position servo system.

Assuming a starting position in which $\theta_0 = \theta_i$, the operation is as follows. If R_1 is adjusted slightly so that U_1 increases, then at first U_2 will not change. The servo amplifier will now be controlled by V_1 and V_2 in such a way that the motor will generate a torque driving the load. The direction must be such that R_2 is adjusted in the same direction as R_1.

The error voltage U_3 will decrease because of this, and after some time we again have for $\theta_0 = \theta_i$, $U_3 = 0$, after which the starting position has been restored.

Let us consider the information at our disposal:

- The turning angle of both potentiometers is 360°; the voltage across the potentiometers is 18 V.
- The gains K_1 and K_2 of V_1 and V_2 are adjustable.
- Between the input of the servo amplifier and the output axle of the servo motor in the normal operating mode, two time constants occur, namely an electrical time constant τ_e (from the field circuit) of 0.06 s, and a mechanical time constant τ_m (from the mass inertia moment and the viscous friction) of 0.16 s. The dc gain of U_5 to n_1 amounts to 120 r.p.m. per volt.
- The tacho generator yields $\frac{1}{80}$ volt per r.p.m.
- $0 \leq \alpha \leq 1$.

In Fig. 8.1(b) we have the block diagram of the total system, according to the above information. The reader may check for himself the correctness of this block diagram by means of the information provided. For this system we shall successively raise some questions and work out the answers.

Question 1:
Suppose $\alpha = 0$ then there is no tacho feedback; $K_2 = 1$; for what value of K_1 will the system start oscillating?

Answer:
The simplified block diagram for this case is shown in Fig. 8.2; the root locus for

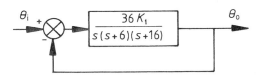

Fig. 8.2 — Block diagram of the system without tacho feedback.

variation of K_1 is sketched in Fig. 8.3.

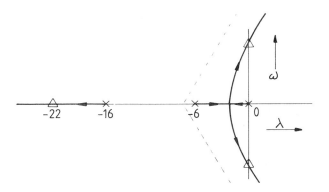

Fig. 8.3 — Root locus of the system without tacho feedback.

The root locus equation is:

$$\frac{36}{s(s+6)(s+16)} = -\frac{1}{K_1} \qquad (8.1)$$

By the summation rule we may determine that oscillation will occur when one of the poles is at $s = -22$; substituting this onto (8.1) results in: $K_{1,\text{osc}} = 58.66$.

Sec. 8.2] A DC position servo system

Question 2:
The tacho feedback remains inoperative; $K_2 = 1$; for what value of K_1 will the settling time (5%) be equal to 2 s, and what will be the overshoot in the step response?

Answer:
The absolute damping needs to be 1.5; then the pole is at $s = -19$ (summation rule). Substituting this pole into the root locus equation (8.1) yields: $K_1 = 20.58$.
    To determine the overshoot we must know the position of the two dominant poles; we don't yet know the imaginary part.
    The root locus equation for $K_1 = 20.58$ may be writen in the form:

$$s^3 + 22s^2 + 96s + 741 = 0 \tag{8.2}$$

Division of the factor $(s + 19)$, for one pole at $s = -19$, yields:

$$s^2 + 3s + 39 = 0 \tag{8.3}$$

From this, the positions of two other poles are $s = -1\tfrac{1}{2} + j6$ and $s = -1\tfrac{1}{2} - j6$.
    The relative damping now becomes: $\lambda/\omega = 1.5/6 = \tfrac{1}{4}$, and the overshoot about 46%.

Question 3:
K_1 retains the value determined in question 2; K_2 and α so that the absolute damping of the system is influenced as favourably as possible.

Answer:
The block diagram for this situation will be as sketched in Fig. 8.4; the root locus

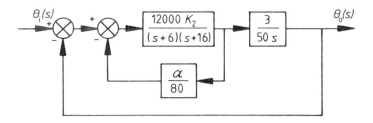

Fig. 8.4 — Block diagram of the system with tacho feedback.

equation for variation of K_2 becomes:

$$\frac{150\alpha s + 720}{s(s+6)(s+16)} = -\frac{1}{K_2} \tag{8.4}$$

Here we see that the tacho feedback introduces a zero. Theoretically the best choice for the position of this zero is the pole $s = -6$. Because of the fluctuation of the poles and zeros that may be expected, $s = -7$ seems to be a recommended choice for the position of the zero; see also section 7.4. Therefore:

$$\frac{720}{150\alpha} = 7, \quad \text{or} \quad \alpha \approx 0.7 \tag{8.5}$$

For this value of α the root locus for variation of K_2 will be as sketched in Fig. 8.5.

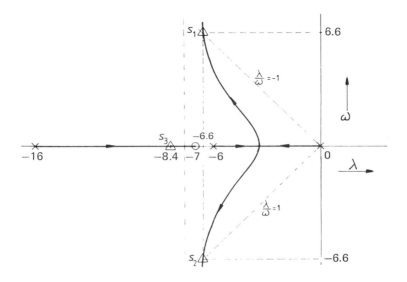

Fig. 8.5 — Root locus for variation of K_2 of the system with tacho feedback.

Question 4:
With the values for K_1 and α that have already been determined, what must be the value of K_2 so that the overshoot will be limited to about 4%?

Answer:
To limit the overshoot to about 4% we must have: $\lambda/\omega = 1$. The poles will then be at:

$$s_1 = \lambda + j\lambda, \quad s_2 = \lambda - j\lambda, \quad \text{and} \quad s_3 = -2\lambda - 22$$

Substituting S_1 into the root locus equation (8.6) yields the following equations:

$$-2\lambda^3 + 96\lambda + 105K_2\lambda + 735K_2 = 0$$
$$2\lambda^3 + 44\lambda^2 + 96\lambda + 105K_2\lambda = 0$$
(8.7)

From (8.7) we may derive the following equation by elimination of K_2:

$$2\lambda^3 + 29\lambda^2 + 154\lambda + 336 = 0 \qquad (8.8)$$

Numerical solution of this equation yields the feasible value of $\lambda = -6.8$, and $K_2 \simeq 1.1$.

With K_1, α, and K_2, the DC position servo system has been set completely. The reader may establish for himself the favourable influence of the tacho feedback on the pole positions by comparing Figs 8.3 and 8.5.

In Fig. 8.6 we see the step responses of the position servo system with and without tacho feedback.

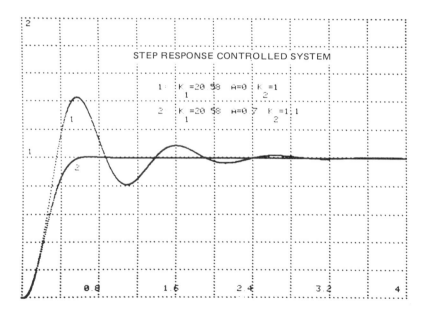

Fig. 8.6—Step responses of the position servo system with (2) and without (1) tacho feedback.

8.3 A PHASE LOCK LOOP

A circuit often found in electronics is the phase lock loop (PLL). The PLL is a feedback control system. It generates an output signal of which the phase is an average value of the phase of the input signal, during a specific time. We shall assume that only sinusoidal signals are produced. Fig. 8.7 is the block diagram of the PLL.

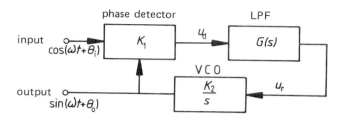

Fig. 8.7 — Block diagram of the PLL.

A phase detector compares the phase of the input signal θ_i to that of the output signal θ_0. The output signal U_d is at first proportional to the phase difference $\theta_i - \theta_0$:

$$u_d = K_1(\theta_i - \theta_0) \tag{8.9}$$

Through a low-pass-filter (LPF) we use this signal as a control signal for a voltage controlled oscillator (VCO). This controllable oscillator yields the ultimate output signal $\sin(\omega t + \theta_0)$. Thanks to the LPF the phase of the output signal does not follow the momentary value of the phase of the input signal, but its average value. The VCO produces a frequency shift $\Delta\omega$ which is proportional to the control signal u_r. Because the phase is the integrated quality of the frequency, the following transfer function of the VCO applies:

$$\frac{\theta_0(s)}{U_r(s)} = \frac{K_2}{s} \tag{8.10}$$

The transfer function of the LPF is $G(s)$. When the LPF ($G(s) = 1$) is not present, the control system represents a feedback integrator. To achieve the correct integration, the time constant of the controlled system must be large. Because of this the bandwidth will be narrow, and consequently the tracking and locking properties (initially and in sudden frequency changes in the input signal) will be poor.

Because it is difficult to find a compromise between the correct integration on the one hand and the correct tracking and locking properties on the other, a filter in the control loop should improve the dymanic behaviour of the above properties.

And there is the important design requirement that the static phase error $\theta_i - \theta_0$ should equal 0. So in any case, the filter should contain an integrator. Therefore the filter must be a PI controller, see Fig. 8.8.

A phase lock loop

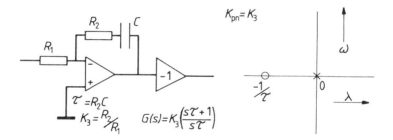

Fig. 8.8 — Feasible realization of a PI controller and the pole/zero plot of $G(s)$.

The transfer function in the control loop now becomes:

$$H_L(s) = \frac{K_1 K_2 K_3 (s\tau + 1)}{s^2 \tau} = K' \frac{(s + 1/\tau)}{s^2} \tag{8.11}$$

and the root locus equation for variation of K_1:

$$\frac{s + \frac{1}{\tau}}{s^2} = -\frac{1}{K'} \tag{8.12}$$

The root locus for $0 < K' < \infty$ is sketched in Fig. 8.9. Usually we want only marginal overshoot in a sudden frequency change, so that $\lambda/\omega = 1$ is a suitable choice for the dominant poles. The relative damping lines will then be at an angle of $\pm 45°$ to the real axis. The poles of the controlled system are indicated by $p^\Delta_{1,2}$. From this the values of K_3 and τ follow.

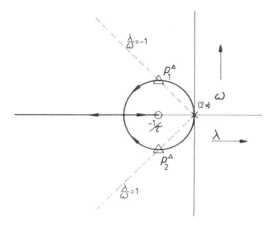

Fig. 8.9 — Root locus of a PI controlled phase lock loop.

For the poles $p_{1,2}^{\Delta}$ of the closed loop system, must the following apply (see root locus):

$$p_{1,2}^{\Delta} = -\frac{1}{\tau}(1 \pm j) \qquad (8.13)$$

From the root locus equation (8.12) we obtain:

$$p_{1,2}^{\Delta} = \frac{-K' \pm \sqrt{K'^2 - 4K'/\tau}}{2} =$$

$$= -\frac{1}{2}\left[K' \pm j\sqrt{\frac{4K'}{\tau} - K'^2}\right] \qquad (8.14)$$

Equating (8.13) and (8.14) yields:

$$K' = 2/\tau \qquad (8.15)$$

The values of τ determines the bandwith of the controlled system. The value of K_1, satisfying the dynamic requirements, is now set by (8.15).

8.4 THE 'GRAVITY RAISER'

In a large class of — mostly mechanical — systems we try to control the position of an object by one or more external forces. As the acceleration of an object is proportional to the resulting external force, according to Newton's law, these systems will contain a double integration, because the position is the second derivative of the acceleration. From preceding chapters it has become quite clear that this is not attractive from a control point of view. (In the frequency domain this means a phase shift of $-180°$!)

The control system discussed here is not a practical example. However, the dynamics of the control is the same as, for example, the tracking system of a rocket.

Moreover, there is an aspect of the stability of the system that is of considerable interest.

In Fig. 8.10 we see that configuration of the system, known by its ambiguous name the 'gravity-raiser'. By 'raising' the gravity by means of a magnetic force, we shall try to float a small metal ball.

It is clear that the condition for floating is that the value of the compensating magnetic force is exactly equal to the gravity force, so controlling the magnetic force is essential. In the realization of Fig. 8.10 this has been done by detecting the position

Sec. 8.4]　　　　　　　　　　　The 'gravity raiser'　　　　　　　　　　　　161

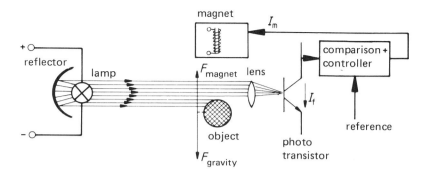

Fig. 8.10 — Floating an object by a magnetic force opposed to gravity.

of the object by a light beam and phototransistor. If the object tends to fall, more light will reach the phototransistor; a solenoid will increase the magnetic force. Moving the object upwards has a reverse effect.

In Fig. 8.11 the block diagram of the control system is shown. Space precludes a

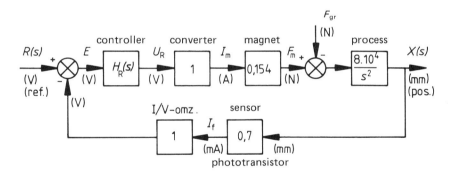

Fig. 8.11 — Block diagram of the control system of Fig. 8.10.

detailed description of the system; stating the various transfer functions obtained from an existing configuration will suffice.

The various transfer functions are valid only in a restricted (linear) field around a certain equilibrium (set point). The following notes on the transfer function of the process are (now) relevant:

1. According to Newton's law:

$$\Sigma F = ma \qquad (8.16)$$

or

$$F_m - F_{gr} = m\frac{d^2x}{dt^2} \qquad (8.17)$$

so that

$$\frac{X(s)}{F_m(s) - F_{gr}(s)} = \frac{1}{ms^2} = \frac{8.10^4}{s^2} \text{ [mm/N]} \qquad (8.18)$$

as the ball weighs 12.5 grams.
2. The magnetic field exhibits a positive feedback; the closer the ball gets to the core of the coil, the stronger the magnetic field will be. This positive feedback is compensated by a position-dependent extra-current negative feedback. These effects are not shown in the block diagram.

In the control of the magnetic field a current control has been applied to eliminate the effect of the extra time constant of the coil (L/R). This time constant has an extra unfavourable effect on the control (extra pole).

For the dynamic behaviour of the control of the system the open loop transfer function $H_L(s)$ is important:

$$H_L(s) = H_R(s) \cdot \frac{8624}{s^2} \qquad (8.19)$$

in which $H_R(s)$ represents the transfer function of the controller. Although the control criteria are rather arbitrary, we assumed the following requirements as to the control behaviour:

a. The steady state error, because of a step disturbance, in the ball's position or set point change must be zero.
b. The damping ratio $\beta \geq 0.7$, or $\lambda/\omega \geq 1$. In connection with the small output area the overshoot in a step must not be too large.
c. The absolute damping λ must be -5 at the least. Within one second the transient phenomena will be reduced to less than 1% variation around the new equilibrium. This requirement corresponds with a settling time t_s (5%) of 0.6 s, or 0.8 s for t_s (2%). Please check!
d. For the current control to function well, the oscillation frequency in the transient phenomenon must not be larger than 25 rad/s. For higher frequencies of the impedance of the coil (ωL) becomes too large, and saturation will occur.

The first requirement — zero steady state error — necessitates an integrating action in the controller, because of the point of impact of the disturbance (F_{gr}); see block diagram. Obviously this only worsens the dynamic behaviour. The other three requirements may be summarized in the position of the poles (of the controlled system) in a specific area in the s-plane; see Fig. 8.12.

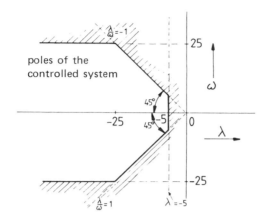

Fig. 8.12 — Requirements for the position of the poles of the controlled system.

Because the process contains a double integration (two-fold pole in the origin), a PI control will result in an unstable system in all circumstances (check this by means of a root locus). Therefore an extra D action is essential. Consequently we choose a (serial-) PID with transfer function $H_R(s)$:

$$H_R(s) = K_R \left(1 + \frac{1}{s\tau_i}\right) \left(\frac{s\tau_d + 1}{0.1s\tau_d + 1}\right) \tag{8.20}$$

So the taming factor in the D action is 10. Thus two zeros ($-1/\tau_i$ and $1/\tau_d$) and two poles (0 and $-10/\tau_d$) are added to the system. The root locus equation is now:

$$\frac{(s\tau_i + 1)(s\tau_d + 1)\,8624}{\tau_i s^3 (0.1s\tau_d + 1)} = -\frac{1}{K_R} \tag{8.21}$$

The root locus for variable K_R is sketched in Fig. 8.13.

The path of the root locus of Fig. 8.13 has an interesting aspect. For values of $K_R \leq K_R'$ the system is unstable. This instability, especially for small values of open loop gain, occurs quite often in this class of control systems.

The poles p_1^Δ to p_4^Δ of the controlled system are indicated in Fig. 8.12. Suitable choices for the most dominant poles are:

$$p_1^\Delta = -5$$

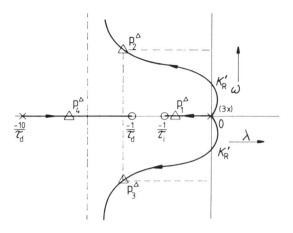

Fig. 8.13 — Root locus of the system with PID control for: $0 < K_R < \infty$.

$$p_{2,3}^\Delta = -25(1 \pm j) \ .$$

The pole $p_{2,3}^\Delta$ is of minor importance. The controller system will have this form:

$$(s+5)(s^2+50s+1250)(s-p_4^\Delta) = 0 \tag{8.22}$$

The root locus equation may be written as:

$$s^4 + \frac{10}{\tau_d} s^3 + 86240 K_R s^2 + 86240 K_R \left(\frac{\tau_i + \tau_d}{\tau_i \tau_d} \right) s +$$

$$+ \frac{86240 K_R}{\tau_i \tau_d} = 0 \tag{8.23}$$

Equalization of equations (8.22) and (8.23) then yields (numerically):

$$K_R = 0.10$$

$$\tau_i = 0.19 \, \text{s}$$

$$\tau_d = 0.056 \, \text{s} \tag{8.24}$$

$$p_4^\Delta = -123.6$$

In Fig. 8.14 the exact root locus has been plotted by means of a computer. From

Sec. 8.4]   The 'gravity raiser'

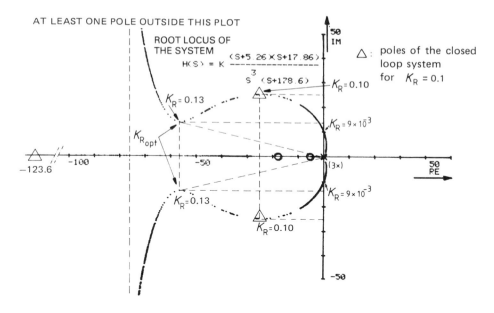

Fig. 8.14 — Root locus plot of the controlled system for $0 < K_R < \infty$, $\tau_i = 0.19\,\text{s}$ and $\tau_d = 0.056\,\text{s}$.

this root locus it appears that the value for $K_R = 0.10$ is not optimum. The optimum value for K_R (assuming $\tau_i = 0.19\,\text{s}$ and $\tau_d = 0.056\,\text{s}$) is indicated in the plot it appears to be about 0.13.

9
Digital computer application in process control

9.1 INTRODUCTION

The systems discussed so far have featured continuity; at any moment the value of all signals was known. Although a great number of control systems show this continuity, there are also many systems which are discontinuous in themselves; only at specific, discrete, instants is the information about the signals in such a system known. In following chapters we shall discuss these sampled data systems or discrete data systems.

An important group of sampled data systems is that in which a digital computer is used. The application of a process computer (a (digital) computer with special features for the automation of processes) has dual aspects. On the one hand such a computer will be used for controlling parts of the process and thus running it in a certain order. On the other hand the computer may be used to control (maintain) specific process parameters. The application of a process computer yields many opportunities for an integral approach to an automation problem with all its control aspects. Usually, such an approach is attended by an increase of quality.

The reliability of such realizations is of great importance in the application of process computers for process automation. This reliability has risen sharply during the last few years. The 'mean time between failures' (MTBF) has risen to often tens of thousands of hours. Of course, this may be achieved only if preventive maintenance is carried out.

The study of sampled data systems will lead to a classification of systems with inherent sampling and of systems in which the sampling is deliberate.

In systems with inherent sampling, the sampling process is a property of the process itself. For example, in a radar installation the information on the object to be surveyed is available only once per revolution of the radar antennae. Another example is the determination of body weight; at regular intervals the body weight is

Sec. 9.1]   Introduction   167

determined by means of scales. For practical reasons it is not usually possible to do this continuously.

In the chemical industry we also find a large number of examples: sampling a chemical, and the analysis of it, takes place at discrete moments. Also in non-technical systems we find examples: economic systems in which information on supplies, etc., is available perhaps only once a month. In systems with deliberate sampling one or more elements are inserted that operate discontinuously. Examples of systems in which the sampling has been inserted deliberately are mostly found in applications of digital techniques.

Digital elements are sequential by nature; that is, the data processing takes some time for each signal magnitude. A discrete time function will then be necessary and will be achieved by sampling.

Applying digital techniques to (usually continuous) processes has certain advantages:

a. Conveying information is accurate and reliable (with minimum noise effects), and storing and processing may be done simply by computer.
b. Digital measurements may often be carried out much more accurately than analog ones. Frequency, r.p.m. and angle measurements (by means of code discs) then rely on the principle that the measurements are in fact a count of zero passages, pulses, and so on, which is a simple and accurate matter.
c. In digital techniques the use of time sharing means a very efficient use of the computer; more systems and processes use the computer simultaneously. By means of multiplexers the control loops are scanned subsequently, and their information is processed.

   During one sampling period (the time between two successive samplings of the same signal) more systems may be scanned. Of course this depends on the processing speed of the computer.

   The maximum number of control loops also depends on the dynamics (speed) of those various sub systems.

   The appearance of inexpensive micro- and mini-computers has decreased the importance of extensive time sharing systems, which are rather vulnerable.

A digital computer controller forming part of the control of a continuous process is shown in Fig. 9.1.

Except for the digital computer itself, we see some 'new' blocks, namely the sample S hold device, the analog-digital converter (ADC), and the digital-analog converter (DAC). These converters form an interface between the digital analog inputs and outputs. As mentioned before it is necessary to make the information discrete in time before computer processing. The sample and hold device (S & H) will sample the analog signal at fixed moments — in Fig. 9.1 by a transducer at the output — and the ADC will make it suitable for the computer by means of digital coding. By means of coding, quantizing occurs. The DA converter does the reverse: the digital code of the computer is converted into an analog signal and made continuous by a hold device (HD).

It is obvious that in accurate operation of the AD and DA converters, coding and decoding respectively affect the accuracy of the total system to a large extent.

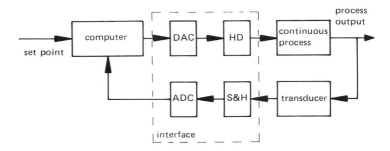

Fig. 9.1 — Digital computer controller.

Because of the overall picture we assume the AD and DA converters to be 'ideal'. We shall not discuss effects of inaccuracy and finite speeds in AD and DA converters. In view of the present developments in techniques this is quite permissible; fast and accurate AD/DA converters are no longer a problem.

However, we must take into account the limited accuracy by which digital words represent signal magnitudes. This accuracy depends on the width or the number of bits (*bi*/nary digi/*ts*) which these words consist of; it is expressed as the resolution of the AD and DA converter. Of course the resolution of the converters must be adapted to the word length of the computer.

Apart from the above advantages of the deliberately sampled data systems there are also disadvantages:

a. The systems are getting more complex and thus more expensive.
b. It will become clear that higher harmonics will develop in the control signal because of the sampling process, by which wear of parts and saturation of amplifiers, etc., may occur.
c. Generally, the stability of the system will decrease. We shall discuss this in detail later. This may be proven by considering that sampling has a time delay effect: only after a complete sampling period may it respond to the next sample; a dead time will develop, which, as we know from the theories of continuous control systems, worsens the stability of the system.

For simplicity we shall consider only systems which are sampled at equidistant moments. Moreover, if there are more samplings in one control loop, it is assumed that these samplings are synchronous.

Signal sampling at equidistant moments is shown in Fig. 9.2. At moments $t = 0, T, 2T, 3T, \ldots$ the signal $x(t)$ is sampled, so that at the output of the sampling device the signal magnitude $x(nT)$ is known at these moments only. The sampling period is T seconds. The sampling frequency will then be:

$$f_s = \frac{1}{T} \text{ Hz.} \tag{9.1}$$

Sec. 9.1] Introduction 169

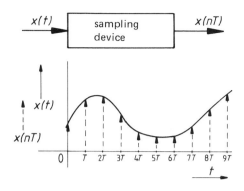

Fig. 9.2 — Signal sampling.

Specific qualities will be given to a process computer used as a special purpose computer, in order to be able to fulfil the tasks given to it. The most important qualities are:

Extensive input/output facilities (see Fig. 9.3)
As in a 'normal' computer we can usually distinguish one flow of data, originating from a one-type input, through the central processor unit (CPU) to the output. In a process computer, however, there are several data flows.

The input and output devices needed interface for the various inputs, and output flows should be sufficiently fast and accurate, and furthermore a flexible programming facility is needed to direct these data flows to and from the CPU.

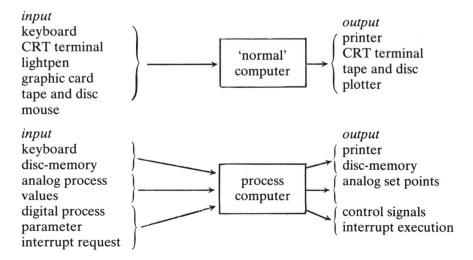

Fig. 9.3 — Computer input and output.

High processing speed
As is usual when controlling a number of control loops consecutively, the processing speed must be high, especially in faster processes with several inputs and outputs. This also implies that the programming must be efficient.

Large memory capacity
To control one or several control loops 'normally' no high demands are made upon the capacity of the internal memory (RAM). For more advanced control algorithms (e.g. optimizing control), and if several tasks must be fulfilled by the process computer, larger memory cpacities are required.

No great demands are made upon a possibly built-in slower disc memory, because here only those data concerning the process may be stored, which need not be rapidly available.

The word length may be limited
As the accuracy with which most physical data of a process are measured, is limited, a word length is 8, 10, 12, or 16 bits may suffice. At 8 bits the resolution error is $100\%/2^8 \simeq 0.4\%$.

Interrupt facilities must be included
When certain limits are exceeded in the process, it must be possible to intervene in order, for example, to take alarm procedures.

A real-time clock is essential
The real-time clock enables the process computer to carry out certain actions at regular moments by interrupting, (e.g. measuring values by analog inputs) or to carry out internal measurements and counts.

In its most simple form, a process computer configuration must consist of:

- a central processor unit
- a data input facility and a data output facility
- one or more memories
- digital inputs and outputs
- analog inputs and outputs
- a real-time clock.

We may conclude that the digital computer will be increasingly important in control engineering or process engineering. On the one hand the computer will more and more be used for analysis, synthesis, and simulation of controlled systems because of the development of ever more versatile software, such as simulation programs. On the other hand the application of computers as part of a control system is becoming more and more common. In chemical industry, in nuclear plants, in steelworks, and in aviation, the computer is becoming an indispensable link in control systems. Often, these systems are so complex, because there are several interlinked inputs and outputs (multivariable systems), that the use of a computer is essential. However, we shall be tempted to realize more systems based on micro-

Sec. 9.2] **'Off-line' computer applications** 171

computers for less complex processes too. The price and flexibility of such systems will become increasingly decisive.

A commonly use of computers in a control system is a distributed control, whereby the control the process itself is done by a so called front-end. This is a microcomputer especially designed for control purposes. A higher level so called supervisory computer, takes care of the trending (data-acquisition), operator interface and parameter setting. The data communication between these computers is mostly serial.

Applying digital computers in a control system may be done in various more or less radical ways. In the next three sections these methods will be discussed in order of increasing integration of process computer and process.

9.2 'OFF-LINE COMPUTER APPLICATIONS

It is clear that the application of a process computer may meet with opposition; for certain 'responsible decisions' will be taken automatically, and very often their reliability is doubted. Other social problem's may arise with the introduction of automation. Especially for existing processes, which up till now have been controlled conventionally, such objections may be overcome only with great difficulty. The computer must prove itself before we turn over more essential tasks to the process computer. Gradual introduction may, if possible, be preferable to introducing a completely automated production process at one go.

At the introduction of process computers there will be a shifting of tasks in the sense that for designing, starting-up, preventive maintenance, and repairs, people with greater skills will be needed. Also, the process operator must be aware that the computer will take over tasks, but that the final responsibility for the process is his.

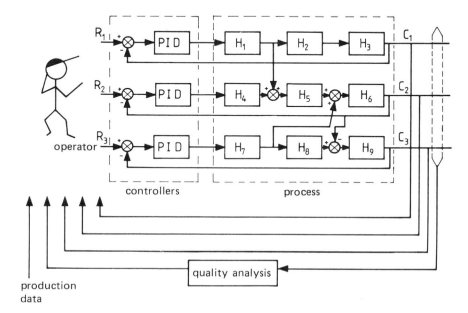

Fig. 9.4 — Controlling a production process.

Let us now consider how controlling a complex process with several control loops is done without a computer, as shown in Fig. 9.4. The process usually consists of a number of variables to be controlled and kept constant. The control loops mostly show interaction through which a change in the set point of one loop influences the value of the other variables. As well as these variables, other data reach the operator, namely the very important data obtained from quality analysis of the product and the desired product. Usually, data from the quality analysis at fixed intervals must be translated by the operator into the right adjustment of the set points. So it is obvious that the experience of the operator is an important factor in more, often interactive, control loops. Thus another operator will very often achieve a different quality of the the product.

In the off-line application of computers there are no direct links between the process computer and the process. All necessary data on the process are put into the computer through human observations and actions. The results of the computer calculations are available to the process operator, who is free to do as he sees fit. It is possible that he decides that some set points must be altered as a result of the data supplied to him by the computer, but he may also ignore them. In fact, the computer advises the operator on the adjustment of the process by means of the data supplied by the operator; see Fig. 9.5. This configuration is different from the usual configuration for controlling a process only by the presence of the process computer, which advises the operator as to the adjustment of the process to be controlled. A large quantity of practical experience and the results of the process may be stored in the process computer, which may be logically interpreted by the process computer and included in its advice to the operator. In this way, the setting of the process will become more independent of changes of operators.

The configuration discussed here is also called the 'off-line operator guide'.

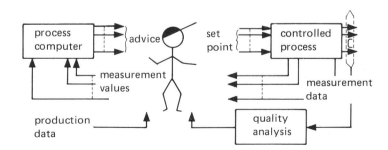

Fig. 9.5 — Off-line computer application.

9.3 ON-LINE APPLICATIONS

In this configuration (Fig. 9.6) there is a direct link between the process computer and the process to be controlled in the sense that measurement data are directly put into the computer. This application opens up the opportunity to carry out the following tasks:

- Regular scanning and measuring of values of process data in order to determine if none of the process values approaches the maximum value or exceeded it. This computer function is also called alarm scanning. If an alarm situation arises, this is made known to the operator, whereupon he should take the necessary measures.
- Regular investigation and recording of a number of process parameters, in order to gather them for reporting purposes. From the recorded data conclusions may very often be drawn for better control of process parameters. This function is often called data logging. The operator is advised as to the adjustment of the various set points of the process.

The computer does not now advise the operator on request, but as soon as appears necessary from the measurement data recorded by the computer itself. The function of the process computer in this configuration is often called on-line operator guidance.

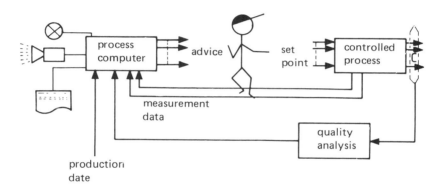

Fig. 9.6 — On-line computer application.

9.4 IN-LINE APPLICATIONS

A logical development of the on-line operator guide configuration is that in which the operator, as a link between computer and process, is replaced by a direct link. The supervision, which before was in the hands of the operator by adjustment of the setpoints of the controlled process, will also be transferred to the computer.

The process-operator has a greater overall supervisory function; he continuously observes the general progress of the production process. In this configuration there are fast control loops which are run by the PID controllers, the process, and the feedback of the controlled values to these controllers. The relatively slow loops are run by the quality analysis and/or the measurement values (through the computer to the set points of the controllers (Fig. 9.7).

When a computer breakdown occurs the process may be kept going by manual adjustment of the set points of the controllers. The level of experience of the operators will still, one hopes, be sufficient to continue the production process.

The configuration discussed is known as 'supervisory control', or 'indirect digital control'.

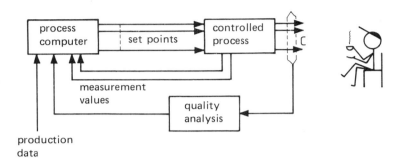

Fig. 9.7 — Supervisory control.

It is also possible for the computer to control the relatively fast loops. Then the configuration is 'direct digital control' or DDC in short; see Fig. 9.8, which features its most simple form. According to this configuration, the operator puts the digital set point data into the computer by means of measurement data.

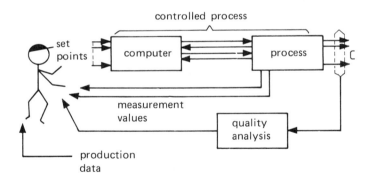

Fig. 9.8 — Direct digital control.

Here the requirements as to the calculation speed of the computer are more severe than in the previous cases; for the computer is now part of the control loops that are directly involved in the process. That is why the calculation speed must be high enough in relation to the relatively fast process dynamics. We shall discuss this in detail later on. Usually, the computer will take over the task of various controllers, in a time sharing regime, so that these requirements for speed will be even more severe. This is also a reason for which one often makes use of the distributed process control mentioned earlier.

Often, we duplicate the task of the computer by back-up controllers, so that in case of breakdown the task of the computer is taken over. This is a back-up system. Sometimes we install only adjusting equipment by which the process is not controlled, but is adjusted in such a way that the process may be kept going for a short period of computer breakdown.

In the direct digital control of Fig. 9.8 the operator still has an important role; among other things he decides the process settings. If this function is also taken over by the computer, we have complete digital control shown in Fig. 9.9.

The computer will keep certain process values constant by means of the fast control loops. The broader measurement data as to the progession of the process, together with the quality analysis and the production requirements, provide the computer with the necessary data to adjust a combination of set points. The operator's task is now a purely supervisory one. But we must not forget that, in such

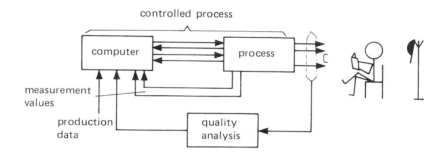

Fig. 9.9 — Complete digital control.

an automatic process, the software is written by people who must know the process very well. Later on, the experience acquired from the process must again and again be incorporated in updating the software.

Because many quality analyses are carried out automatedly, in some cases the communication between quality analysis equipment and process computer need no longer go via the operator, but may be carried out directly by the process computer.

It is also possible that the decision as to the product to be manufactured and the quality are taken by means of a computer. For example on the basis of the raw materials supplied this higher level computer will pass on the necessary data directly to one or more process computers. In this case we may speak of a hierarchy of computers in a specific production system.

9.5 PROBLEMS

a. In a production process we usually speak of steering and controlling. Give examples of both aspects.
b. What is the difference between inherent and deliberate signal sampling?

c. Mention advantages and disadvantages of computer aided control.
d. What are the requirements for a process computer?
e. What is meant by off-line operator guide?
f. What is meant by on-line operator guide?
g. What is meant by supervisory control?
h. What is meant by DDC?
i. Describe the configuration of a controlled process with complete digital control.
j. What will be the consequences, if there is a computer breakdown in the configurations mentioned in (e) through (i)?

10
Direct digital control (DDC)

10.1 INTRODUCTION

As we have seen in Chapter 9, in direct digital control the computer is included in the relatively fast control loops. The control actions are stored in the computer in the form of control algorithms, by means of which new control signals are determined by one or more measurement values and the desired values. The signals occurring in such loops are multifarious; we may find both continuous and sampled signals, and both analog and discrete digitally coded signals.

The components occurring in the control loop in this form of controlling will be discussed in this chapter, and the influence of these components on control behaviour will be examined. We shall emphasize a qualitative approach rather than an extact mathematical description of the occurring phenomena, which will be discussed in the later chapters.

10.2 THE CONTROL LOOP

Fig. 10.1 shows a direct digital control for keeping constant three controlled values C_1, C_2 and C_3. It operates as follows. In the sensors the values C_1 to C_3 are adapted to the sample and hold device (SH). These devices, comprise field effect transistors T_1 to T_6, the holding capacitors C_1 to C_3, and the buffer amplifiers B_1 to B_3. They sample the measurement signals U_1, U_2, U_3 and keep this sample at an (almost) constant value during the conversion time of the analog-digital converter. Normally, the signals G_1, G_2, and G_3 are low; the p-channel transistors T_1, T_2 and T_3 will then conduct, so that the capacitor voltages U_{c_1}, U_{c_2}, and U_{c_3} will follow the voltages U_1, U_2 and U_3 respectively. If we now wish input 2 to be sampled, then voltage G_2 is increased. Transistor T_3 will now block, and T_5 (n-channel) will conduct. Capacitor C_2 will keep almost the voltage it had just before the blocking of T_2, while via T_5 this voltage is supplied to the input of the ADC. As soon as the conversion is completed, the signal CD will become high; the code representing U_2 may then be read into the

178   Direct digital control (DDC)   [Ch. 10

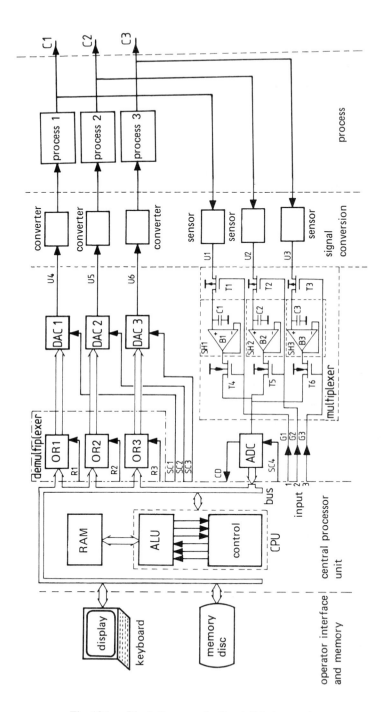

Fig. 10 1 — Block diagram of a direct digital control.

CPU. After comparing the code indicating the value of U_2 to the value desired, the error signal is put into the control algorithm and a new control magnitude for process 2 is calculated. As soon as the calculation of this control magnitude is completed, the output register OR_2 is activated. This will take the newly calculated control magnitude from the data bus. Then the Digital Analog Converter DAC_2 is addressed to indicate that conversion may commence. The voltage U_5 will then assume the value calculated by the CPU as a new control signal. Via signal adaptation 2 — e.g. an amplification of voltage current conversion — the process no. 2 may be controlled such that the error signal is decreased.

The other loops operate likewise. New set point values may be put in by means of the keyboard, while measurement values and alarm messages may be read out on the display.

It is clear that different signal forms occur in the loop, by which the system behaviour may distinguish it from non-digital controlled systems. The part of the control loop in which these different signals occur as shown in Fig. 10.2.

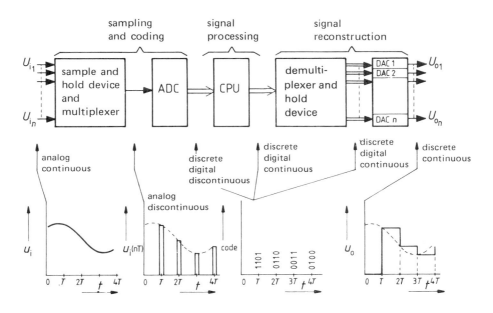

Fig. 10.2 — Signal forms in the control loop.

The multiplexers, which make sure that a number of control loops may be connected to one process computer, are not essential here, in the sense that they do not influence the consequences of signal sampling and signal reconstruction for the control loop. It is obvious that the analog input signal is discreted both in time and in amplitude. The discretion in time develops because the signal value is determined at certain instants and is held in the form of an amplitide of a pulse. The discretion in the amplitude value is converted into a binary coded number. As we shall see in a stadium, both discretions influence the behaviour of the control loop. Thus the number of times that discretion occurs in time (in other words the sampling frequency) influences the dynamic properties of the control loop, while the accuracy of the quantitizing (the level of accuracy with which the number reflects the amplitude) has a direct effect on the steady state accuracy of the control loop.

10.3 INTERFACES

To realize the control loops a number of circuits must be added to the computer:

- An analog multiplexer, with which a choice is made from the supplied input signals. In one loop the multiplexer is, of course, left out.
- A sample and hold device with which a sample is taken from the input signal and its value is kept constant for a short time.
- An analog-digital converter which converts the analog value of the sample into a digital code.
- A digital (de-)multiplexer with which the value of the newly calculated control signal is supplied to a certain output.
- A digital-analog converter which converts the supplied digital code into a (discrete) voltage.

Because of the analog multiplexer we need only one ADC; the latter is relatively expensive. The construction of such a multiplexer appears from the (simplified) block diagram in Fig. 10.1, and consists of transistors T_1 to T_6. This multiplexer is programmable; that is the inputs may be selected at random.

The sample and hold device must be able to track the input signal properly. Therefore the time constant in the input circuit must be small with regard to the time constants of the process. Usually the sensor also contains a filter by which the noise is eliminated from the signal. The time constant by which the capacitor is discharged must be large, so that the condensor voltage will stay almost constant during the hold period. It is usually necessary that the ADC is preceded by an anti-aliasing filter, to prevent high-frequency appearing in the sampling signal as low-frequency components. (See also Chapter 17).

The input formed by the multiplexer and the sample and hold device must satisfy the following requirements:

- There must be a galvanic decoupling between the actual input and the converter. An optical link, by means of a light-emitting diode (LED) and a phototransistor, may be used here.
- The input must affect the actual measurement signal as little as possible; that is, the input must be of high impedance.
- Common signals at both connections of one input must be suppressed in the converters; only the differential signal is important. This effect is expressed in the

common mode rejection ratio (CMRR), this is the degree to which a differential voltage has more influence than a common voltage.

The analog-(digital) converter transfers the output voltage of the SH into a binary code. Some important qualities of the ADC are:

- The conversion speed
  The speed of the conversion depends on the chosen conversion ssytem and the number of bits in which the signal value is expressed after conversion.
- The resolution
  The renumber of bits of the number that denotes the magnitude of the sampled is called the resolution.
  The number of bits determines the deviation between the real signal value and the represented binary number. The resolution is mostly expressed in bits, but may also be in percentages. Thus a 10-bit resolution means that, at most, the conversion error will be 1/1024 ($2^{10} = 1024$) of the maximum amplitude on the ADC, and this may also be indicated by a resolution of 0.1%.
- The overall accuracy
  As well as the resolution, there are some other factors that determine the overall accuracy, e.g. accuracy and stability of reference voltages, sensitivity to temperature variations, etc. The overall accuracy is less than might be suggested by the number of bits. Modern ADCs have an inaccuracy of twice the value of the least significant bit (LSB), also indicated by ± LSB.

Two frequently occurring analog-(digital) conversion principles will now be discussed, namely the successive approximation and the dual slope method.

Successive approximation ADC
As suggested by its name, this ADC operates in successive approximations to determine the digital code. The diagram for a 4-bit code is given in Fig. 10.3. The operation is as follows:

In the logic circuit, half the maximum analog input voltage is generated as first reference value for the analog input voltage, in the form of the binary number 1000 (8). This value is converted by the DAC into an analog voltage and is compared with the input voltage by the comparator. If the analog input voltage U_a is higher than the reference voltage U_b, we have $p = 1$, $q = 0$ and $r = 1$. The logic circuit is now so activated by q, that at the next clock pulse the reference value is increased by half to 1100 (12), after which a new comparison takes place.

At this clock pulse the signal $r = 1$ was put into the output register as MSB of the result of the conversion.

If, after reference, $q = 1$ ($U_a < U_b$), the new reference value is decreased by half the preceding step. After four references and five clock pulse the binary number representing the result of the conversion is in the output register, with the MSB at the right and the LSB at the left.

Let us assume that the maximum voltage that may be converted is 15 V and the analog input voltage is 13.5 V; then conversion proceeds by the next steps:

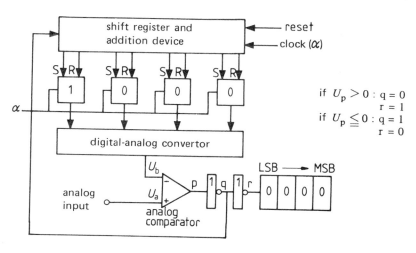

Fig. 10.3 — Principle of a successive approximation ADC.

Clock pulse	Comparison register	Comparator	p	q	r	Output register
1	1 0 0 0 (8)	$U_a > U_b$	1	0	1	0 0 0 0
2	1 1 0 0 (12)	$U_a > U_b$	1	0	1	1 0 0 0
3	1 1 1 0 (14)	$U_a < U_b$	0	1	0	1 1 0 0
4	1 1 0 1 (13)	$U_a > U_b$	1	0	1	0 1 1 0
5	0 0 0 0	$U_a > U_b$	1	0	1	1 0 1 1

The binary number representing the analog input voltage U_a after conversion will then be 1101 (13 decimal). The necessary conversion time is about 1 µs per bit; for a 10-bit ADC this means that a maximum of 10^5 conversions per second may be carried out. It must be observed that time is needed for the various CPU instructions too, so that the number of conversions will be less. However, for control matters this conversion speed is sufficient.

Dual-slope ADC
This dual slope ADC operates on the principle of an integration of the analog signal (first slope) followed by a discharge of the integration capacitor (second slope). In Fig. 10.4 this is shown in a principle diagram and a time diagram. The operation is as follows:

At $t = t_0$ a start conversion impulse is given with which also the binary counter is reset by the logic circuit. The analog input voltage $U_{a,1}$ that must be converted is connected to the integrator by transistor T_1, by which the voltage U_b becomes negative if U_a is positive. The integration of U_a takes until $t = t_1$, by which the interval

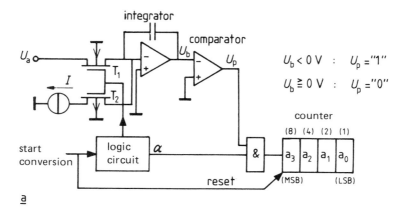

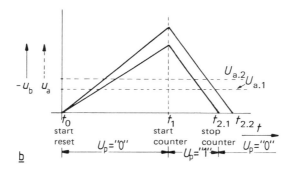

Fig. 10.4 — Principle of a dual-slope ADC.

$t_0 - t_1$ is fixed. At $t = t_1$, T_1 is blocked and T_2 becomes conductive, so that the integrator is connected to the current source I, which discharges the integration capacitor by a given slope. From $t = t_1$ the clock pulse α is also supplied to the AND port; because U_b is negative $U_p = $ '1', and the clock pulse is supplied to the counter. Thus the counter increases itself until at $t = t_{2,1}$ $U_b = 0$ V and thus $U_p = 0$, and the clock pulse is blocked. At that moment the counter reading is a measure of the analog input voltage. If $U_a > U_{a,1}$, for example $U_2 = U_{a,2}$, then the final integration voltage will be higher and the discharge will take proportionally longer. Also, the final counter reading will be higher. Fluctuations around the average value of U_a are averaged by the integration, which is an advantage of this type of ADC. A disadvantage may be the relatively long conversion time. If t_2 is 50 ms at maximum, the number of conversions will be 20 per second at maximum.

The ADC code is binary; that is, the base is two, with only figures 0 and 1. For positive and negative values of U_a the code must also indicate the polarity. Usually the bit on the left is the polarity bit, in which 0 represents positive (+) and 1 negative

(−). For positive values almost all codes are the same; for negative values some different codes are used, namely signed magnitude, one's complement and two's complement. The code may also be offset-binary. For more information on the usual codes we refer to the available literature.

The digital (de-)modulator by which a certain value is assigned to a certain channel consists of a number of output registers with a common data bus. By making it accessible by means of an 'enable'-signal, one of the output registers may be filled with new information on the control signal of the process concerned.

Each output channel is supplied with a DAC which converts the code stored in the matching output register in a discrete voltage value.

So long as an output register has a certain code, the DAC will issue the matching voltage. The function of the output register is also that of a hold device by which the output voltage becomes continuous. (See also section 10.6).

The operation of a DAC is mostly based on weighed currents; these are currents of which the value is proportional to the significance of a specific bit of the input code. Thus the LSB with significance 1 will connect half the current of the bit with significance two, etc. In Fig. 10.5 a principle diagram of a 4-bit DAC is shown; the

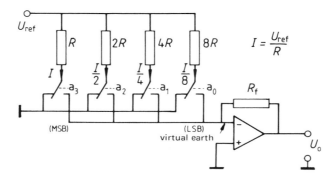

Fig. 10.5 — Principle of a current switching DAC.

operation is as follows. The switches — of course in reality these are electronic switches — which are controlled by a_0 through a_3, connect the current to earth if these signals are 0 and to the input of the summing amplifier if these signals are 1. This input behaves like virtual earth; that is, the input voltage is almost always zero. For the output voltage we have:

$$U_0 = -\left\{\frac{a_3}{R} + \frac{a_2}{2R} + \frac{a_1}{4R} + \frac{a_0}{8R}\right\} U_{ref} \cdot R_f \quad (10.1)$$

in which a_0 to a_3 are 0 or 1. If the reference voltage U_{ref} is chosen to be negative, U_0 becomes positive.

It is obvious that the output voltage cannot assume all voltage values but only a number of discrete values; this number is 2^n if n is the number of bits of the input code. Also, a DAC is subject to all sorts of errors, such as inaccuracies and non-linearities, which will not be discussed further.

10.4 SIGNAL SAMPLING

The sampling theorem of Shannon is an important factor in the sampling of signals. According to this theorem no information will be lost in a sampling frequency (f_s) which is equal to or higher than twice the highest frequency f occurring in the signal to be sampled (see also section 11.4), so:

$$f_s \geq 2 f_h \qquad (10.2)$$

Example 10.1
A measure signal has a bandwidth of 1000 Hz. An adequate sampling arises when at least 2000 equidistant samples per second are taken from this signal. Because of loss of information in further processing we usually take a larger margin, e.g.:

$$f_s \geq (5 \ldots 10) f_h \qquad (10.3)$$

The sampling process may be realized in different ways; some methods are indicated in Fig. 10.6.

Fig. 10.6a shows the result of a sampling in which the input signal appears at the output of the circuit at every time interval T during a time τ. This may happen either by means of a switch which is closed at each interval T during time τ, or by means of a modulator. In the latter case the modulating signal has the form shown in the diagram.

Fig. 10.6b shows the method by which a sample and hold device is applied. Note that now the output signal is constant during the interval τ, while in the two methods described above the input signal is tracked.

Thus the output signal of the sampling circuit $U^*(t)$ consists of a series of pulses of width τ; the magnitude of these pulses is converted by the ADC into a binary number. We write the result, after coding, as $U^*(nT)$ and feed this signal into the CPU of the computer.

10.5 SIGNAL PROCESSING IN THE CPU

The CPU receives discrete signals in the form of binary words and issues discrete signals of the same form. Between input and output there is a signal processing procedure which may be described in the most general sense as follows:

$$a_0 U_a^*(0) + a_1 U_a^*(-T) + \ldots + a_{n-1} U_a^*(-nT+T) + a_n U_a^*(-nT) =$$
$$b_0 U_0^*(0) + b_1 U_0^*(-T) + \ldots + b_{n-1} U_0^*(-nT+T) + b_n U_0^*(-nT) \qquad (10.4)$$

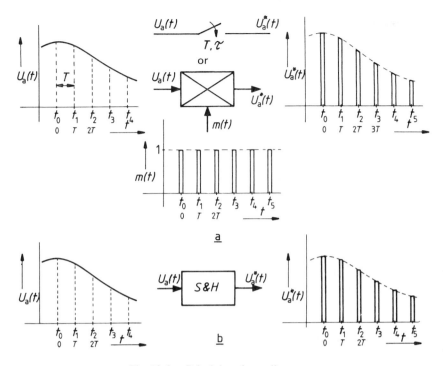

Fig. 10.6 — Principles of sampling.

Here $U_a^*(0)$ is the momentary input signal, $U_a^*(-T)$ the preceding input signal, and so on. The factors a_i and b_i determine the degree to which the signal values in (10.4) are taken into account. We may also write expression (10.4) in the form from which it will be much clearer how the momentary value of the output signal is calculated:

$$U_0^*(0) = \frac{a_o}{b_0} U_a^*(0) + \frac{a_1}{b_0} U_a^*(-T) + \ldots + \frac{a_n}{b_0} U_a^*(-nT) +$$

$$- \frac{b_1}{b_0} U_0^*(-T) - \frac{b_2}{b_0} U_0^*(-2T) - \ldots - \frac{b_n}{b_0} U_0^*(-nT) \qquad (10.5)$$

It is obvious that we cannot take into account the future values of input and output signals.

The meaning of formulas (10.4) and (10.5) will be dealt with in Chapter 14, when discussing control algorithms.

Example 10.2
To calculate the momentary output value of the CPU we use:

$$U_0^*(0) = 4\, U_a^*(0) + 2\, U_a^*(-T) - 0.1\, U_a^*(-2T) + 0.5\, U_0^*(-T) \qquad (10.6)$$

The calculation of the momentary output signal may be represented schematically as in Fig. 10.7.

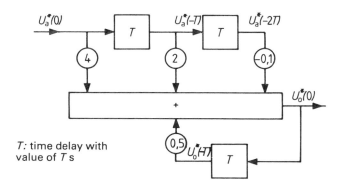

T: time delay with value of T s

Fig. 10.7 — Calculation diagram of the CPU.

If, for example, $U_a^*(0) = 2$, $U_a^*(-T) = 5$, $U_a^*(-2T) = 3$, and $U_0^*(-T) = 10$, the new output value will be $U_0^*(0) = 22.7$.

Each next sampling is put into the CPU whereby also the stored values of the necessary preceding input and output signals are replaced by the next, etc. In section 14.2 we shall discuss in detail the form of a calculation diagram in Fig. 10.7.

10.6 SIGNAL RECONSTRUCTION

If the binary code of the output signal of the CPU is fed to the DAC during a time τ then a pulse signal will develop at the output of the DAC. If the sampling period of the input signal (T) is much longer than the time delay in the CPU and the other hardware, then every T seconds a pulse will develop at the output of the DAC. The width of the pulse will be considered more closely; and is determined, without further provisions, by the duration of the supply of the code word by the CPU to the DAC.

There are several disadvantages in feeding such a pulse signal to the system to be controlled:

- Many higher harmonics with relatively large amplitudes will arise because of the pulse form by which deterioration and saturation effects may occur. However, we must note that most systems show a low-pass character so that the effect of the higher harmonics is decreased. In servomechanisms, however, there may be increased deterioration in e.g. the gearbox.
- The short pulse signal contains but little energy to control the system. A possibility might be to increase the amplitude of the pulse, but this may promote the above mentioned effects. A better solution would be to widen the pulse; this will be discussed below.

- If only the pulse is applied, control information is fed to the process at the sampling instants only. It is much better to feed control information to the process in the intervening time also, by interpolation or extrapolation of some of the sampled signal values.

The disadvantages mentioned above have led to such reconstruction of the control signal that it becomes continuous and will not contain too many higher harmonics with large amplitudes. For this filters are necessary; a practical realisation is found by means of hold devices. Both methods of reconstruction will be discussed.

According to Shannon's sampling theorem no information is lost in sampling an analog signal if the sampling frequency is at least twice as high as the highest frequency occurring in the analog signal. If this requirement is satisfied the analog signal may be completely reconstructed by means of the following formula:

$$f(0t) = \sum_{n=0}^{2\omega T} F\left(\frac{n}{2\omega}\right) \frac{\sin 2\pi\omega\left(\frac{t-n}{2\omega}\right)}{2\pi\omega\left(t - \frac{n}{2\omega}\right)} \quad (10.7)$$

where $F\left(\dfrac{n}{2\omega}\right)$: is the sampling value at time $t = \dfrac{n}{2\omega}$.

ω : is the highest frequency occurring in the analog input signal.

T : is the duration of the analog input signal (T must be large).

$2\omega T$: is the total number of samples.

$\dfrac{\sin p}{p}$: with $p = 2\pi\omega\left(t - \dfrac{n}{2\omega}\right)$ is a cardinal function of the form given in Fig. 10.8.

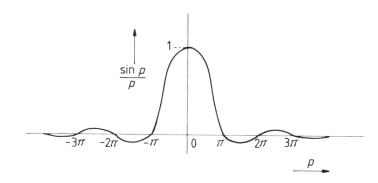

Fig. 10.8 — Cardinal function for the Shannon filter.

Sec. 10.6]   Signal reconstruction

Thus the reconstruction according to the Shannon process starts from the fact that a cardinal function is made up of every sampling pulse with the same amplitude at $p=0$ as this sampling pulse. Moreover, all sampling values, including future values, should be taken into account in the summation.

It is clear that this method is not practically realizable, and moreover is not desirable because of the phase lag in such a filter.

A simpler reconstruction may take place by filtering the sampling impulse in such a way that, as a result of the pulse, a voltage form arises at the output of the filter as shown in Fig. 10.9. This solution is an approximation of the cardinal function according to Shannon's reconstruction.

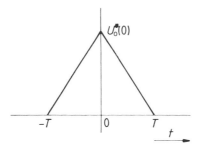

Fig. 10.9 — Triangle approximation of the cardinal function.

By summation of two half triangles, where the height of one of them has the value of the preceding steering pulse ($t = -T$) and the other one a height proportional to the momentary pulse ($t = 0$), a linear interpolation arises between two consequtive signal values. However, it is clear that the reconstructed control signal may only be reconstructed from $t = -T$ to $t = 0$ after the signal value at $t = 0$ has been released by the ADC. This means an extra delay of T seconds due to the reconstruction process.

A simpler form of reconstruction is achieved if we apply hold devices, whereby the signal is reconstructed by means of extrapolation methods between the last sampling value and the next (still unknown) sampling value.

In the time domain the signal reconstruction of $x^*(t)$ may be understood to be an extrapolation of the signal between nT and $(n+1)T$, if the values of $x^*(t)$ at the preceding moments are known. For a continuous signal, $x(t)$, between $t = nT$ and $t = (n+1)T$, Taylor's expansion gives:

$$x(t) = x(nT) + (t - nT)\frac{dx}{dt}\bigg|_{t=nT} + \frac{(t-nT)^2}{2!}\frac{d^2x}{dt^2}\bigg|_{t=nT} + \ldots \qquad (10.8)$$

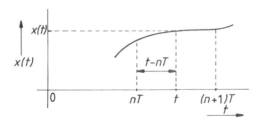

Fig. 10.10 — Extrapolation of $x(t)$ between nT and $(n+1)T$.

Formula (10.8) is an infinite series; the more terms of this series may be determined for increased accuracy of the value of $x(t)$. It will be obvious that this also increases the technical problems of realization.

The number of terms of formula (10.8) determines the order of the process, which functions as an extrapolation. The simplest and also most widely used extrapolator is the zero order hold (ZOH), given by:

$$x(t) = x(nT) \text{ and } nT < t < (n+1)T \tag{10.9}$$

The value of $x(t)$ at $t = nT$ is held for the duration of the sampling period.

An extrapolator that is a little more accurate is the first order hold, given by:

$$x(t) = x(nT) + (t - nT)\left.\frac{dx}{dt}\right|_{t=nT} \text{ and } nT < t < (n+1)T$$

For short sample time T (thus $f_{sample} \gg f_{signal}$) this relation changes into:

$$x(t) = x(nT) + (t - nT)\frac{x(nT) - x(nT - T)}{T} \tag{10.10}$$

A zero order hold (ZOH) takes into account only the signal value at $t = nT$ for the reconstruction of the signal in the interval $nT \leq t < (n+1)T$ and keeps it constant throughout this interval; for some examples see Fig. 10.11. From these examples it appears that the average phase lag due to the reconstruction is about T; see also section 10.7. It is clear that such a ZOH is well able to reconstruct a sampled constant signal. A uniformly increasing or decreasing sampled signal is reconstructed from the sampling values as a regularly increasing or decreasing series of steps.

A first order hold (FOH) reconstructs a continuous output signal by means of the momentary sampling value and a preceding one. The reconstruction is done according to formula (10.10). The effect of the reconstruction is that, starting from the sampling value at $t = nT$ in the interval $nT < t < nT + T$, a signal is generated that

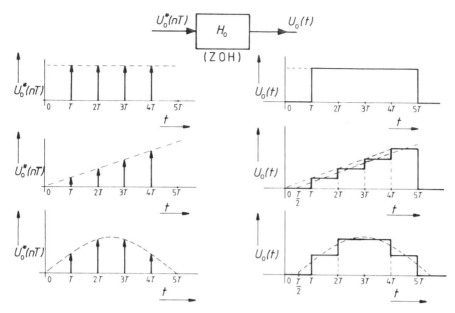

Fig. 10.11 — Some examples of signal reconstruction by ZOH.

increases or decreases according to the slope between the sampling values at $t = (n-1)T$ and $t = nT$. There are some examples in Fig. 10.12.

A ZOH is especially suitable if the sampling values are constant; uniformly increasing signals are reconstructed well by a FOH without too many input transient disturbances.

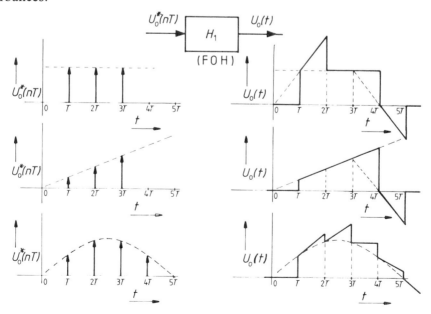

Fig. 10.12 — Some examples of signal reconstruction by FOH.

10.7 QUALITATIVE CONSIDERATION OF SAMPLING AND RECONSTRUCTION

The effect of a hold device on sampling values originating from various analog signal forms has been described in Chapter 9. We will now take a more detailed look at the properties of zero order hold, which occurs in almost every control system.

First we determine the transfer function $G_{ho}(s)$ whereby we start from the impulse response of such a ZOH; see Fig. 10.13.

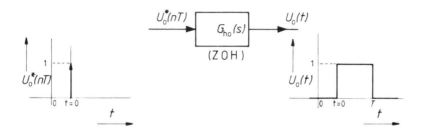

Fig. 10.13 — Determination of the transfer function of a ZOH.

The input pulse, originating from sampling a signal amplitude 1, has an energy content of 1. This is indicated by an arrow in Fig. 10.13. This will be discussed in detail later on. We may assume the output signal (see Fig. 10.13) to be constructed from a unit step at $t=0$ followed by an identical but negative step at $t=T$. This becomes:

$$G_{ho}(s) = \frac{U_0(s)}{U_0^*(s)} = \frac{\frac{1}{s} - \frac{e^{-sT}}{s}}{1} = \frac{1-e^{-sT}}{s} \qquad (10.11)$$

To understand the ZOH in the frequency domain, we substitute $s = j\omega$ in (10.11):

$$G_{ho}(j\omega) = \frac{1-e^{-j\omega T}}{j\omega} \qquad (10.12)$$

Sec. 10.7] Qualitative consideration of sampling and reconstruction

And also:

$$\begin{aligned}
G_{ho}(j\omega) &= \frac{e^{-\frac{1}{2}j\omega T}(e^{-\frac{1}{2}j\omega T} - e^{-\frac{1}{2}j\omega T})}{j\omega} \\
&= \frac{e^{-\frac{1}{2}j\omega T} \cdot 2j \sin\frac{1}{2}\omega T}{j\omega} \\
&= \frac{T \sin\frac{1}{2}\omega T \cdot e^{-\frac{1}{2}j\omega T}}{\frac{1}{2}\omega T}
\end{aligned} \qquad (10.13)$$

So for the modulus and the argument of this transfer function, we have:

$$|G_{ho}(j\omega)| = \frac{T \sin\frac{1}{2}\omega T}{\frac{1}{2}\omega T} \qquad (10.14)$$

$$\varphi = \arg\{G_{ho}(j\omega)\} = \arg\left\{\frac{\sin\frac{1}{2}\omega T}{\frac{1}{2}\omega T}\right\} - \frac{1}{2}\omega T \qquad (10.15)$$

The Bode diagram of $G_{ho}(j\omega)$ is sketched in Fig. 10.14.

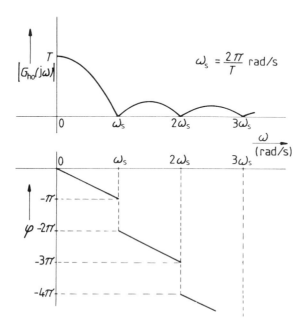

Fig. 10.14 — Bode diagram of $G_{ho}(j\omega)$.

Note that:

1. $G_{ho}(j\omega)$ is zero for: $\frac{1}{2}\omega T = k\pi$ or for $\omega = \frac{k\pi}{\frac{1}{2}T} = k\frac{2\pi}{T} = k\omega_s$, with $k = 1,2,3 \ldots$

2. $\lim_{\omega \to 0} \frac{\sin\frac{1}{2}\omega T}{\frac{1}{2}\omega T} = 1$, so that $|G_{ho}(jo)| = T$.

3. In the phase characteristic the phase leaps repeatedly by $-\pi$ at multiples of ω_s. This is caused by the sign changes in the term $\frac{\sin\frac{1}{2}\omega T}{\frac{1}{2}\omega T}$.

4. The scales in this Bode diagram are all linear.

5. The filtering operation of such a ZOH is obvious.

To determine the frequency spectrum of a sampled signal we consider the situation of Fig. 10.15a. If the input signal is sinusoidal with frequency μ and the

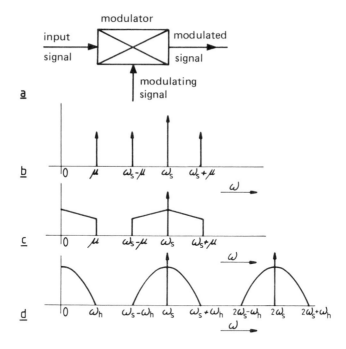

Fig. 10.15 — Examples of frequency spectra of a modulated signal.

modulating signal is a sinusoidal carrier with frequency ω_s, then the modulated signal also contains components with frequencies $\omega_s \pm \mu$ in addition to the original frequency μ. In this case Fig.10.15 shows the frequency spectrum of the output signal. If the input signal covers the complete frequency band from 0 to μ, then the frequency spectrum of Fig. 10.15c arises.

Sec. 10.7]   **Qualitative consideration of sampling and reconstruction**   195

The sampling process may be understood to be the multiplication of the input signal by an impulse series, which contains, except for the base frequency, all higher harmonics. If the sample frequency is ω_s and the maximum frequency in the input signal is ω_h, then we find the frequency spectrum for this case in Fig. 10.15d.

The effect of reconstruction by means of a ZOH is shown in Fig. 10.16 by means

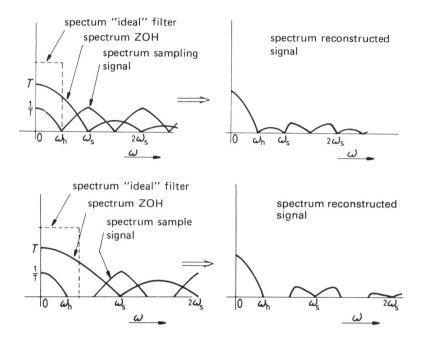

Fig. 10.16 — Frequency spectra at different sampling frequencies.

of frequency spectra. Fig. 10.16a shows the situation if the highest signal frequency $\omega_h = \frac{1}{2}\omega_s$ (minimum theoretical sampling frequency according to Shannon), and Fig. 10.16b shows the situation for a higher sampling frequency. The dotted line in Fig. 10.16 indicates the frequency characteristic of the 'ideal' filter.

From the figure it becomes clear that by taking a higher sampling frequency, fewer high-frequency components arise in the reconstructed signal. Consequently the filtering qualities of the ZOH will improve.

From Fig. 10.16 it also appears that if $\omega_h = \frac{1}{2}\omega_s$ no satisfying filtering will be obtained by a ZOH.

In practice therefore, we often choose $\omega_s \approx 5\text{–}10\omega_h$. Besides, the 'ideal' filter may only be approximated if the order of the hold *proceeds to infinity*.

From the Bode diagram in Fig. 10.14 it is clear that for frequencies that are much lower than ω_s the phase lag is proportional to ω. Then $\varphi = -\omega T_d$, and for $\omega = \omega_s$, $\varphi = -\pi$. Thus:

$$T_d = \frac{T}{2} \qquad (10.16)$$

The result of (10.16) may also be understood directly from (10.13). Here we see that for $\omega_h \ll \omega_s$ — as always in practice — $|G_{ho}(j\omega)|$ is approximately constant and has a value T.

As we shall see in Chapter 11, an attenuation factor $1/T$ arises in the sampling process, so that on the whole no signal strength is lost if the sampling is followed by reconstruction. Viewed in this light the effect of signal sampling and signal reconstruction may be taken into account by a ZOH by a time delay of T seconds. This approximation is only valid if $\omega_h \ll \omega_s$ and if the signal processing in the CPU takes little time compared to the sampling period T. See also formula (10.13). An often applied method of system analysis of sampled data systems is that in which sample and hold devices are replaced by $\frac{1}{2}T$ time delays. Then conventional control theories are applied.

10.8 PROBLEMS

1. Draw a block diagram of a DDC control loop, name the components, and describe the nature of the various signals.
2. Important parameters in a DDC loop are the sampling frequency and the resolution of ADC and DAC. What control properties are influenced by these parameters respectively?
3. What is required of the time constant of the hold device of a sample and hold with respect to tracking and holding the input voltage?
4. Name some important properties in an ADC.
5. Describe the principle of an ADC by means of successive approximation and the dual slope method respectively.
6. An 8-bit ADC samples and codes sinusoidal signals $x(t) = A \sin \omega t$. The amplitude A is 10 V maximum. The sampling time needed to determine the value of the input signal with sufficient accuracy, is 10 μs.

    Determine the maximum frequency of the input signal, if the error due to the infinite sampling time lies within the limits of the LSB of the converter.

    Note: Assume dx/dt to be constant for the duration of the sampling.
7. Describe the principles of a current switching DAC.
8. What is Shannon's sampling theorem?
9. A signal processing in the CPU with input signal $U_a^*(nT)$ and output signal $U_0^*(nT)$ is described by:

$$4 U_a^*(0) + 2 U_a^*(-T) = 7 U_0^*(0) + 8 U_0^*(-T) \qquad (10.17)$$

where: $U_a^*(-T) = -1$, $U_a^*(0) = 0$, $U_a^*(T) = 1$, and $U_0^*(-T) = 2$.
Find: $U_0^*(0)$ and $U_0^*(T)$.

Also draw the calculation diagram.
10. Let us assume the sampling data of Fig. 10.17. Draw the output signal of a ZOH

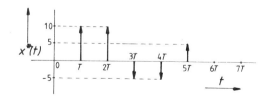

Fig. 10.17 — Sampling data of Problem 10.

and a FOH, if the pulses of Fig. 10.17 are fed to the hold device.
11. Which approximation for the effects of sampling and ZOH construction is commonly used?
12. Let us assume a sampled data system as shown in Fig. 10.18.

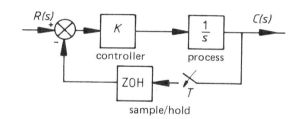

Fig. 10.18 — The sampled data system according to Problem 12.

Determine roughly the positive values of K (as a function of T) for which the system is stable.

Note:
Remember that the system becomes unstable if the total open loop gain is 1 and the phase is $-\pi$.

11

Mathematical description of sampled data systems

11.1 INTRODUCTION

In this chapter we shall first discuss some methods for describing discrete (sampled) signals, from which will appear the specific character of such signals. However, the disadvantages of these description methods are so great that we shall use a description that is obtained from a suitable transform, the z-transform. As a result it will become possible to consider a system in the z-domain which closely corresponds to the considerations of continuous signals and systems in the s-domain, discussed in the first few chapters of this book. The properties and rules applied there are also valid in the z-domain, which provides an extra stimulus to apply the z-transform.

11.2 DESCRIPTION OF A SAMPLED DATA SYSTEM BY MEANS OF AN IMPULSE SEQUENCE

So far we have considered the sampling process to be a switch closing every T seconds for a duration τ. To simplify the mathematical description considerably we assume 'ideal sampling', i.e. the 'switch' that takes care of the sampling closes infinitely fast. Yet, to give the pulses of the sampled signal some energy content, we must assume the amplitude of the pulses to be infinitely great, so that the areas will equal one. For the determination of the transfer function $G_{ho}(S)$ of the ZOH have we already used this method; see formula (10.11). Of course this is not so in practical devices; the switch closes for a short time τ during which the ADC has the opportunity to recognize the signal magnitude. This is shown in Fig. 11.1. The sampling period is T.

The sampled signal is represented by $x_\tau^*(t)$. Our first assumption is that the switch closes for so short a time that the sampled signal will be constant during the closing time (this is the flat-top approximation). Therefore:

Description of a sampled data system

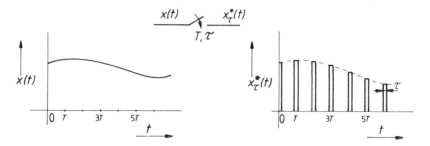

Fig. 11.1 — Switch with finite closing time.

$$x^*_\tau(t) = x(kT) \text{ for } kT \leq t \leq kT + \tau$$

and $\quad x^*_\tau(t) = 0 \quad$ for $kT + \tau < t < (k+1)T$, with:
$$k = 0, 1, 2, 3, \ldots \tag{11.1}$$

So:

$$x^*_\tau(t) = \sum_{k=0}^{\infty} x(kT) \left[1\,(t - kT) - 1\,(t - kT - \tau) \right] \tag{11.2}$$

in which

$$1\,(t - a) = 1 \text{ for } t > a$$
$$= 0 \text{ for } t < a \quad \text{(Heaviside function)}$$

Applying the Laplace transform to expression (11.2) yields:

$$X^*_\tau(s) = \sum_{k=0}^{\infty} x(kT) \left[\frac{e^{-ksT}}{s} - \frac{e^{-ksT} \cdot e^{-s\tau}}{s} \right] = \sum_{k=0}^{\infty} x(kT)\,[1 - e^{-s\tau}] \cdot \frac{e^{-skT}}{s}$$

Because τ is small we may assume:

$$1 - e^{-s\tau} = 1 - \left(1 - s\tau + \frac{s^2 \tau^2}{2!} - \frac{s^3 \tau^3}{3!} + \ldots\right) \simeq s\tau \tag{11.3}$$

From (11.2) and (11.3) follows:

$$X^*_\tau(s) \simeq \sum_{k=0}^{\infty} x(kT)\, s\tau\, \frac{e^{-skT}}{s} = \tau \sum_{\kappa=0}^{\infty} x(kT)\, e^{-ksT} \tag{11.4}$$

After reverse transform to the time domain we obtain:

$$x_\tau^*(t) = \tau \sum_{k=0}^{\infty} x(kT)\cdot\delta(t-kT) = \tau x^*(t) \tag{11.5}$$

In 11.5, $x^*(t)$ represents the sampling magnitudes series, written as a series of impulses with energy contents equalling the sampling magnitudes. This is indicated in Fig. 11.2.

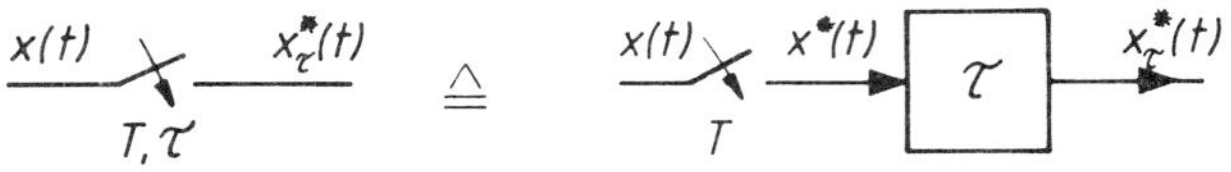

Fig. 11.2 — Equivalence of a 'practical' and an 'ideal' sampling device.

From this figure it is also clear that a non-ideal sampling device may be replaced by an ideal one followed by a factor τ. So this approximation is allowed only if τ is small with regard to the sampling time.

Because a hold is incorporated in almost every sampled data system, the sampled signal will, as it were, be 'spread' over a complete sampling period. In this way the factor τ is compensated again. On account of this we assume $\tau = 1$ for convenience, so that in mathematical processing we are able to work with an ideal sampling device if $\tau \ll T$.

If there is no hold device in the system, this is not allowed, and formula (11.5) must be applied.

As indicated above, sampling may be seen as periodically closing (periodicity T) a switch during a short time τ or as a modulation process. As the result of both methods is the same, the mathematical description will also be the same.

The result of such sampling of a signal $x(t)$ is written as $x^*(t)$. This sampled signal consists of a series of pulses of width τ and an amplitude of $x(nT)$. If $\tau \ll T$ and the sampling at least satisfies Shannon's theorem (see expressions (10.2) and (10.3)), then the amplitudes of $x(nT)$ will be almost constant.

The pulse series may be written as:

$$x^*(t) = [x(nT)], \text{ with } n = 0, 1, 2, \ldots$$
$$= x(0), x(T), x(2T). \ldots \tag{11.6}$$

The width τ of the pulses is determined in practice by the construction of the input circuit, formed by the sample and hold and the ADC. On the one hand τ must not be too large, for then the amplitude of a pulse varies too much during sampling; on the

Description of a sampled data system

other hand τ must not be too small because then the energy content of the pulse becomes too small. The pulse amplitude not being constant, may sometimes be disadvantageous to the accuracy of the analog-digital conversion. Too narrow a pulse with small energy content would be influenced too much by the load circuit with an RC input character.

However, theoretically the description of the sampling process with finite pulse-width τ is rather difficult. If the pulse width is very narrow compared to the sampling period T, it is possible to describe the sampled signal by an impulse series. Such an impulse series may be obtained by multiplying the analog input signal of the sampling device by a unit impulse series.

The unit impulse series consists of a series of pulses at times $t = 0, 2T, 3T, \ldots$ with pulse width τ and pulse amplitude $1/\tau$, in which $\tau \rightarrow 0$; see also section 10.7. Such pulses are also indicated as Dirac or δ pulses. The energy content fits the area of the pulses and thus $\tau \cdot \frac{1}{\tau} = 1$. Such a pulse series may be written as:

$$\delta_T(t) = [\delta(nT)] = \sum_{n=0}^{\infty} \delta(t - nT) \tag{11.7}$$

For the sampled signal $x^*(t)$ we may also use the modulator theory:

$$x^*(t) = x(t)\delta_T(t) = x(t) \cdot \sum_{n=0}^{\infty} \delta(t - nT) \tag{11.8}$$

Because $\delta_T(t) = 0$ for $t \neq 0, T, 2T, 3T, \ldots$ and assuming that $x(t) = 0$ for $t < 0$, formula (11.8) becomes:

$$x^*(t) = \sum_{n=0}^{\infty} x(nT) \cdot \delta(t - nT) \tag{11.9}$$

and

$$x^*(t) = x(0)\delta(t) + x(T)\delta(t - T) + x(2T)\delta(t - 2T) + \ldots \tag{11.10}$$

So then sampled signal consists of a series of pulses with energy content $x(nT)$ and originates from a sampling device that is assumed to be 'ideal'.

Example 11.1
The analog signal that must be sampled has the form:

$$x(t) = 4 \text{ e}^{-2t} \tag{11.11}$$

The sampling period is $\tfrac{1}{2}s$. The impulse series describing the sampled signal $x^*(t)$ will be (according to formula (11.10)):

$$x^*(t) = 4\delta(t) + 1.47\delta(t - \tfrac{1}{2}) + 0.54\delta(t - 1) + 0.2\delta(t - 1\tfrac{1}{2}) + \ldots \tag{11.12}$$

See also Fig. 11.3.

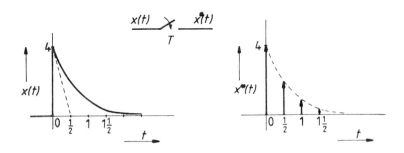

Fig. 11.3 — A sampled signal indicated by a pulse series.

11.3 THE LAPLACE TRANSFORM OF SAMPLED DATA

The Laplace transform $X^*(s)$ of $x^*(t)$ may be determined as follows:

$$X^*(s) = \mathcal{L}\{x^*(t)\} = \mathcal{L}\left\{\sum_{n=0}^{\infty} x(nT)\delta(t - nT)\right\} =$$

$$= \int_0^{\infty} \sum_{n=0}^{\infty} x(nT)\delta(t - nT) \, \text{e}^{-st} \, dt \tag{11.13}$$

Note:
To determine the Laplace transform the following generally applies:

$$\mathcal{L}\{f(t)\} = \int_0^{\infty} f(t) \text{e}^{-st} \, dt. \tag{11.14}$$

Expression (11.13) only makes sense at times $t = 0, T, 2T, 3T, \ldots$, so for $X^*(s)$ we may write:

The Laplace transform of sampled data

$$X^*(s) = \int_0^\infty \sum_{n=0}^\infty x(nT)\delta(t-nT)e^{-snT}dt =$$

$$= \int_0^\infty x(0)\delta(t)dt + \int_0^\infty x(T)\delta(t-T)\,e^{-sT}\,dt +$$

$$+ \int_0^\infty x(2T)\,\delta(t-2T\,e^{-2sT}\,dt + \ldots =$$

$$= x(0)\int_0^\infty \delta(t)\,dt + x(T)\,e^{-sT}\int_0^\infty \delta(t-T)\,dt +$$

$$+ x(2T)\,e^{-2sT}\int_0^\infty \delta(t-2T)\,dt + \ldots \tag{11.15}$$

Generally, the integral of a unit impulse equals 1, for the area is 1. That is why expression (11.15) becomes:

$$X^*(s) = x(0) + x(T)\,e^{-sT} + x(2T)\,e^{-2sT} + \ldots \tag{11.16}$$

and

$$X^*(s) = \sum_{n=0}^\infty x(nT)\,e^{-snT} \tag{11.17}$$

By means of formula (11.17) it is, in principle, possible to determine the Laplace transform of every sampled signal, provided that the Laplace transform of the original, not yet sampled, signal exists.

Example 11.2
Assume that $x(t) = e^{-at}$. Determine $X^*(s)$.

Solution:
(see expression (11.17))

$$X^*(s) = \sum_{n=0}^\infty x(nT)\,e^{-snT} = \sum_{n=0}^\infty e^{-anT} \cdot e^{-snT} = \sum_{n=0}^\infty e^{-(s+a)nT} =$$
$$= 1 + e^{-(s+a)T} + e^{-2(s+a)T} + e^{-3(s+a)T} + \ldots \tag{11.18}$$

Expression (11.18) represents a geometrical series with base 1 and ratio $e^{-(s+a)T}$, so that we may also write:

$$X^*(s) = \frac{1}{1-e^{-(s+a)T}} = \frac{1}{1-e^{-sT}\,e^{-aT}} \tag{11.19}$$

This simple example shows us that the expression for $X^*(s)$ is totally different from the one for $X(s)$; for the latter equals $X(s) = \dfrac{1}{s+a}$. The meaning and the use of Laplace-transforms of sampled data will be discussed later.

11.4 FOURIER SERIES OF SAMPLED DATA

For analysis in the frequency domain the sampled form can be developed in a Fourier series. According to (11.8) the sampled form of a signal $x(T)$ may be written as

$$x^*(t) = x(t) \sum_{n=0}^{\infty} \delta(t - nT)$$

First we may now develop the unit impulse series $\delta_T(t) = \sum_{n=0}^{\infty} \delta(t - nT)$ in a Fourier series. Therefore we first consider this impulse series with pulse width τ and pulse amplitude $1/\tau$; so the area is 1. From this we determine the Fourier series, and after that we let the pulse width be zero. In Fig. 11.4 the pulse series with pulse width τ is shown.

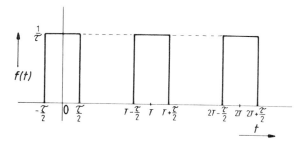

Fig. 11.4 — Pulse series with finite dimensions.

The general formula to determine the Fourier series is:

$$f(t) = a_0 + \sum_{n=1}^{\infty} a_n \cos n\omega t + \sum_{n=1}^{\infty} b_n \sin n\omega t, \qquad (11.20)$$

with

Fourier series of sampled data

$$a_0 = \frac{1}{T}\int_0^T f(t)\, dt$$

$$a_n = \frac{2}{T}\int_0^T f(t)\cos n\omega t\, dt$$

$$b_0 = \frac{2}{T}\int_0^T f(t)\sin n\omega t\, dt$$

The function $f(t)$ is symmetric with respect to the Y-axis, therefore $b_n = 0$. For a_0, the following applies:

$$a_0 = \frac{1}{T}\int_0^T f(t)dt = \frac{1}{T}\left[\int_0^{\frac{1}{2}\tau}\frac{1}{\tau}dt + \int_{T-\frac{1}{2}\tau}^T \frac{1}{\tau}dt\right] = \frac{1}{T}\left[\frac{1}{2}+\frac{1}{2}\right] = \frac{1}{T} \quad (11.21)$$

For a_n the following applies:

$$a_n = \frac{2}{T}\int_0^T f(t)\cos n\omega t\, dt = \frac{2}{T}\int_0^{\frac{1}{2}\tau}\frac{1}{\tau}\cos n\omega t\, dt + \frac{2}{T}\int_{T-\frac{1}{2}\tau}^T \frac{1}{\tau}\cos n\omega t\, dt$$

For reasons of symmetry we may write this as:

$$a_n = 2\cdot\frac{2}{T}\int_0^{\frac{1}{2}\tau}\frac{1}{\tau}\cos n\omega t\, dt = \frac{4}{\tau T}\int_0^{\frac{1}{2}\tau}\cos n\omega t\, dt = \frac{4}{n\omega\tau T}\sin n\omega t \Big|_0^{\frac{1}{2}\tau} = \frac{4\sin\frac{1}{2}n\omega\tau}{Tn\omega\tau} \quad (11.22)$$

If the limit for $\tau \to 0$ we obtain:

$$\lim_{\tau\to 0}\frac{2}{T}\cdot\frac{\sin\frac{1}{2}n\omega\tau}{\frac{1}{2}n\omega\tau} = \frac{2}{T} \quad (11.23)$$

By means of (11.20), (11.21), and (11.23) we have for $\sum_{n=0}^{\infty}\delta(t-nT)$:

$$\sum_{n=0}^{\infty} \delta(t-nT) = \frac{1}{T} + \frac{2}{T} \sum_{n=1}^{\infty} \cos n\omega t =$$

$$= \frac{1}{T} [1 + 2 \cos \omega t + 2 \cos 2\omega t + 2 \cos 3\omega t + \ldots]$$

(11.24)

Application of Euler's theorem ($e^{j\phi} = \cos \phi + j \sin \phi$) gives:

$$\sum_{n=0}^{\infty} \delta(t-nT) = \frac{1}{T} [1 + e^{j\omega t} + e^{-j\omega t} + e^{2j\omega t} + e^{-2j\omega t} + \ldots] =$$

$$= \frac{1}{T} \sum_{n=-\infty}^{\infty} e^{jn\omega t} = \frac{1}{T} \sum_{n=-\infty}^{\infty} e^{-jn\omega t}$$

(11.25)

in which $\omega = \omega_s$ $2\pi/T$ is the sampling frequency (in rad/s). Substitution of (11.25) in (11.8) renders for the sampled signal $x^*(t)$:

$$x^*(t) = x(t) \frac{1}{T} \sum_{n=-\infty}^{\infty} e^{-jn\omega_s t}$$

(11.26)

The Laplace transform, $X^*(s)$, is:

$$X^*(s) = \mathcal{L}\, [x^*(t)] = \int_0^{\infty} x(t) \frac{1}{T} \sum_{n=-\infty}^{\infty} e^{jn\omega_s t} \cdot e^{-st} \, dt =$$

$$= \frac{1}{T} \sum_{n=-\infty}^{\infty} \int_0^{\infty} x(t)\, e^{-jn\omega_s t}\, e^{-st}\, dt = \frac{1}{T} \sum_{n=-\infty}^{\infty} \int_0^{\infty} x(t)\, e^{-(s+jn\omega_s)t}$$

or:

$$X^*(s) = \frac{1}{T} \sum_{n=-\infty}^{\infty} X(s + jn\omega_s)$$

(11.27)

The meaning of formula (11.27) is explained by the following example:

Example 11.3
$x(t) = e^{-at}$; thus $X(s) = 1/(s+a)$. By means of (11.27) $X^*(s)$ becomes:

Sec. 11.4] Fourier series of sampled data

$$X^*(s) = \frac{1}{T} \sum_{n=-\infty}^{\infty} \frac{1}{s+a+jn\omega_s} =$$

$$= \frac{1}{T} \left[\frac{1}{s+a} + \frac{1}{s+a+j\omega_s} + \frac{1}{s+a-j\omega_s} + \cdots \right] =$$

$$= \frac{1}{T} \left[\frac{T(s)}{(s+a)(s+a+j\omega_s)(s+a-j\omega_s)(s+a+2j\omega_s)(s+a-2j\omega_s)\cdots} \right] \quad (11.28)$$

Fig. 11.5 shows the pole representation of $X^*(s)$.

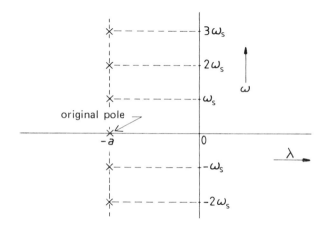

Fig. 11.5 — Pole plot of $X^*(s)$ if $X(s) = 1/(s+a)$.

From this pole plot it appears that the original pole $s = a$ of $X(s)$ is repeated every other ω_s. So the number of poles of $X^*(s)$ is infinitely large. It is considerably more difficult to determine the position of the zeros of $X^*(s)$, but this is not important for our further considerations.

According to formula (11.27) we may generally assume that the pole plot of any Laplace transformed form $X^*(s)$, whether or not $X(s)$ describes a signal or a system, shows the periodicity found. In Fig. 11.6 this is shown for a system with 3 poles (s_1, s_2, s_3). The part of the s-plane between $+\frac{1}{2}j\omega_s$ and $-\frac{1}{2}j\omega_s$ is called the primary strip.

The frequency spectrum of the sampled signal may easily be determined by substitution $s = j\omega$ into (11.27). Then we find:

$$X^*(j\omega) = \frac{1}{T} \sum_{n=-\infty}^{\infty} X(j\omega + jn\omega_s) \quad (11.29)$$

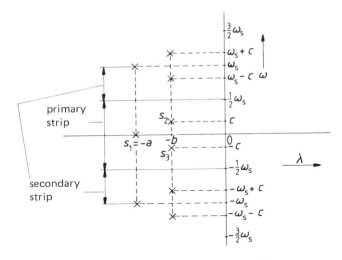

Fig. 11.6 — Pole plot of $X^*(s)$ if $X(s) = \dfrac{1}{(s+a)(s^2+2bs+b^2+c^2)}$.

From formula (11.29) it appears that if the original signal contains a frequency component ω_1, the sampled signal will contain the frequencies $\omega_1 \pm n\omega_s$. The frequency components are higher harmonics. Of course the same may be said for a complete frequency band. The frequency band (the spectrum) will then be repeated at a frequency ω_s; see also section 10.7. Thereby infinitely many high frequent spectra will arise. This is shown in Fig. 11.7, by means of a given frequency spectrum (modulus characteristic) of $X(j\omega)$. In this figure ω_h represents the highest occurring frequency in the spectrum of $X(j\omega)$. From this figure it also appears that for $\omega_h = \tfrac{1}{2}\omega_s$ the frequency bands exactly meet. Then it is still possible to retrieve $X(j\omega)$ by means of an ideal filter (dashed line) from the sampled signal. If $\omega_h > \tfrac{1}{2}\omega_s$, it will not be possible to filter the original signal from the sampled signal without distortion. See Fig. 11.8.

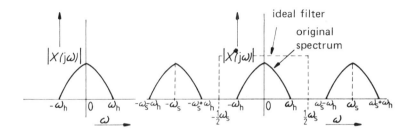

Fig. 11.7 — Frequency spectrum of $X^*(j\omega)$ at given $X(j\omega)$.

Fourier series of sampled data

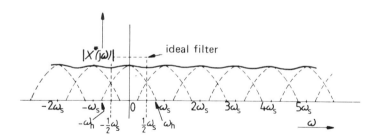

Fig. 11.8 — Frequency spectrum of $X^*(j\omega)$ if $\omega_h > \tfrac{1}{2}\omega_s$.

The sampling requirement following from the above:

$$\omega_s \geq 2\omega_h, \text{ or: } f_s \geq 2f_h \qquad (11.30)$$

is Shannon's sampling theorem (see section 10.4). If this requirement is not satisfied, as indicated in, for example, Fig. 11.8, the frequency bands will show aliasing, and the high frequency components of $X(j\omega)$ may no longer be distinguished from the sampling process.

If a signal of 1 Hz is sampled with 3 Hz, the original signal may be reconstructed. If, however, the sample frequency is decreased to $\tfrac{4}{3}$ Hz, then a signal of $\tfrac{1}{3}$ Hz will develop in reconstruction.

In general, we must make sure that, if the sampling frequency cannot be chosen sufficiently high to prevent aliasing, the high frequency components in the signal to be sampled should be filtered out before sampling. The filters applied are called anti-aliasing-filters.

Note that the factor $1/T$ in formula (11.29) is compensated by the factor T in formula 10.14, so that no signal strength will be lost in sampling and reconstruction.

Summarizing, we may say that we have found two expressions for $X^*(s)$, one by direct application of the Laplace transform and the other by the Fourier series. The latter is very suitable for analysis in the frequency domain by substitution of $s = j\omega$.

Direct method:

$$X^*(s) = \sum_{n=0}^{\infty} x(nT)\, e^{-snT}$$

Fourier series:

$$X^*(s) = \frac{1}{T} \sum_{n=-\infty}^{\infty} X(s + jn\omega_s)$$

Frequency domain method:

$$X^*(j\omega) = \frac{1}{T} \sum_{n=-\infty}^{\infty} X(j\omega + jn\omega_s)$$

11.5 THE z-TRANSFORM

In Example 11.2 we have seen that the Laplace transform of a sampled signal is totally different from that of a discrete signal. We shall see that the mathematical considerations of sampled data systems will be simplified if we utilize the so-called z-transform. In this transform we apply the following substitution to a sampled expression:

$$e^{-sT} = z^{-1} \tag{11.31}$$

or:

$$s = \frac{1}{T} \ln z \tag{11.32}$$

Regarding later applications of this transform it is important that we understand that factor z^{-1} in fact comes down to a time delay equal to the sampling period T. (See expression (11.31)). By means of the substitution introduced in formula (11.32), we may write for the Laplace transformed form $X^*(s)$ of $x^*(t)$, according to (11.17):

$$X(z) = X^*(s)\bigg|_{s=\frac{1}{T}\ln z} = \sum_{n=0}^{\infty} x(nT)\, e^{-snT}\bigg|_{s=\frac{1}{T}\ln z}$$

or

$$X(z) = \sum_{n=0}^{\infty} x(nT) z^{-n} \tag{11.33}$$

$$X(z) = x(0) + x(T)z^{-1} + x(2T)z^{-2} + \ldots \tag{11.34}$$

Except for sampled signals, it is also possible, as will appear later, to apply the z-transform to series of numbers as well as to elements that process discrete signals, the discrete elements.

In the next four examples the z-transform will be applied according to (11.33) or

(11.34). We shall see that the z-transforms often represent convergent goemetric series as in Examples 11.4, 11.5, and 11.6. In Example 11.7 the z-transformed form of a random series of numbers is determined.

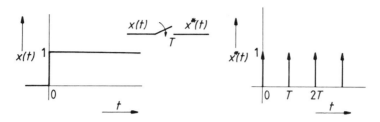

Fig. 11.9 — Sampling a unit step.

Example 11.4
Determination of the z-transform of a sampled unit step; see Fig. 11.9.
We have:

$$x(t) = 1(t); \text{ that is, } x(t) = 0 \text{ for } t < 0$$
$$\text{and } x(t) = 1 \text{ for } t > 0.$$

So:

$$x^*(t) = \sum_{n=0}^{\infty} x(nT)\delta(t - nT) \text{ with } x(nT) = 1.$$

According to formula (11.34), $X(z)$ is:

$$X(z) = \sum_{n=0}^{\infty} x(nT)z^{-n} = 1 + z^{-1} + z^{-2} + \ldots$$

This expression represents a geometric series of which $X(z)$ is the sum, and is expressed by:

$$X(z) = \frac{1}{1 - z^{-1}} = \frac{z}{z - 1} \tag{11.35}$$

Example 11.5
Determination of the z-transform of the sampled signal of $x(t) = e^{-at}$.
For $x^*(t)$ we have:

$$x^*(t) = \sum_{n=0}^{\infty} e^{-anT} \delta(t - nT)$$

so that according to formulas (11.33) and (11.34), $X(z)$ is given by:

$$X(z) = \sum_{n=0}^{\infty} x(nT)z^{-n} = \sum_{n=0}^{\infty} e^{-anT} z^{-n} = 1 + e^{-aT}z^{-1} + e^{-2aT}z^{-2} + \ldots$$

Then for $X(z)$ we find, as in Example 11.4:

$$X(z) = \frac{1}{1 - e^{-aT} z^{-1}} = \frac{z}{z - e^{-aT}} \qquad (11.36)$$

Substitution of $e^{-sT} = z^{-1}$ into the result of Example (11.2) yields the same answer, of course. Please check!

Example 11.6
Determination of the z-transform of $x(nT) = a^n$.
   The series $[x(nT)]$ may be written as:

$$[x(nT)] = 1, a, a^2, a^3 \ldots$$

This becomes:

$$X(z) = \sum_{n=0}^{\infty} x(nT)z^{-n} = \sum_{n=0}^{\infty} a^n z^{-n} = 1 + az^{-1} + a^2 z^{-2} + \ldots =$$

$$= \frac{1}{1 - az^{-1}} = \frac{z}{z - a} \qquad (11.37)$$

Example 11.7
Determination of the z-transform of a random series of numbers. Given the series: $[x(nT)] = 3, 1, 0, -2, 0, 0, 0$, the z-transform of $[x(nT)]$ will be:

$$X(z) = \sum_{n=0}^{\infty} x(nT)z^{-n} = 3 + 1z^{-1} + 0z^{-2} + (-2)z^{-3} = 3 + \frac{1}{z} - \frac{2}{z^3} =$$
$$= \frac{3z^3 + z^2 - 2}{z^3} \tag{11.38}$$

11.6 THE TABLE OF z-TRANSFORMS AND SOME PROPERTIES OF THE z-TRANSFORM

By means of expression (11.33) z-transforms may be determined. In Table 11.1 the z-transforms of most function occurring in control engineering are given. In this table we also find some properties of the z-transform. For evidence on these properties we refer to more extensive literature on this subject. Unless otherwise stated all signals start at $t=0$.

The z-transforms in Table 11.1 have been determined for the sampled expression, thus for $x^*(t)$ or for $X^*(s)$. So wherever $F(z)$, in this book, is directly determined from $F(s)$, we mean the z-transform of the sampled signal $f^*(t)$. Of course this is only possible if $f^*(t)$ exists.

Example 11.8
Given:

$$X(s) = \frac{1}{s(s+1)} \tag{11.39}$$

what is $X(z)$?

Solution:
We shall write $X(s)$ in such a form that the elements of $X(s)$ may be recognised in the table. Note that $\mathscr{L}\{f_1(s) \cdot f_2(s)\} = F_1(z) \cdot F_2(z)$ does not apply. By partial fraction of $X(s)$ a suitable form for transforms will develop, for:

$$X(s) = \frac{1}{s(s+1)} = \frac{1}{s} - \frac{1}{s+1}$$

From this we obtain, regarding property 10:

$$\mathscr{L}\{X(s)\} = \mathscr{L}\left\{\frac{1}{s} - \frac{1}{s+1}\right\} = \mathscr{L}\left\{\frac{1}{s}\right\} - \mathscr{L}\left\{\frac{1}{s+1}\right\}$$

Rules 1 and 3 from Table 11.1 now yield:

$$X(z) = \frac{z}{z-1} - \frac{z}{z-e^{-T}} = \frac{z(1-e^{-T})}{(z-1) \cdot (z-e^{-T})} \tag{11.40}$$

Table 11.1 — Table of z-transforms and properties

	System	signal $f(t)$	$F(s)$	$F(z)$
1	Integrator	1	$\dfrac{1}{s}$	$\dfrac{z}{z-1}$
2	Two integrators	t	$\dfrac{1}{s^2}$	$\dfrac{zT}{(z-1)^2}$
3	First order system	e^{-at}	$\dfrac{1}{s+a}$	$\dfrac{z}{z-e^{-aT}}$
4	—	te^{-at}	$\dfrac{1}{(s+a)^2}$	$\dfrac{zTe^{-aT}}{(z-e^{-aT})^2}$
5	Undamped second order system	$\sin bt$	$\dfrac{b}{s^2+b^2}$	$\dfrac{z\sin bT}{z^2-2z\cos bT+1}$
		$\cos bt$	$\dfrac{s}{s^2+b^2}$	$\dfrac{z(z-\cos bT)}{z^2-2z\cos bT+1}$
6	Damped second order system	$e^{-at}\sin bt$	$\dfrac{b}{(s+a)^2+b^2}$	$\dfrac{ze^{-aT}\sin bT}{z^2-2ze^{-aT}\cos bT+e^{-2aT}}$
		$e^{-at}\cos bt$	$\dfrac{s+a}{(s+a)^2+b^2}$	$\dfrac{z^2-ze^{-aT}\cos bT}{z^2-2ze^{-aT}\cos bT+e^{-2aT}}$
7	—	$x(nT)=a^n$	—	$\dfrac{z}{z-a}$

Properties

8 Linearity rule

$$af_1(t) \pm bf_2(t) \Leftrightarrow aF_1(z) \pm bF_2(z)$$

9 Shift rule

$$f(t-mT) \Leftrightarrow z^{-m} F(z)$$

10 Multiplying by t

$$t \cdot f(t) \Leftrightarrow -Tz\frac{d}{dz}\{F(z)\}$$

11 Limit value theorems

a $\lim\limits_{t \to 0} f(t) = \lim\limits_{z \to \infty} F(z)$ b $\lim\limits_{t \to \infty} f(t) = \lim\limits_{z \to 1} \dfrac{z-1}{z} F(z)$

Example 11.9
Given

$$X(s) = \frac{e^{-sT_d}}{s+2}, \text{ with } T_d = 2T. \tag{11.41}$$

what is $X(z)$?

Solution:
According to rule 9,

$$X(z) = \mathcal{L}\left\{\frac{e^{-2sT}}{s+2}\right\} = z^{-2}\mathcal{L}\left\{\frac{1}{s+2}\right\} = z^{-2} \cdot \frac{z}{z-e^{-2T}} = \frac{1}{z(z-e^{-2T})} \tag{11.42}$$

Example 11.10
Given

$$x(t) = 4(1-t)\,e^{-2t} \tag{11.43}$$

what is $X(z)$.

Solution:

$$x(t) = 4(e^{-2t} - te^{-2t})$$

According to rules 3 and 4 we now have:

$$X(z) = 4\left\{\frac{z}{z-e^{-2T}} - \frac{zTe^{-2T}}{(z-e^{-2T})^2}\right\} = 4\,\frac{z(z-e^{-2T}(1+T))}{(z-e^{-2T})^2} \tag{11.44}$$

11.7 INVERSE z-TRANSFORM
For inverse transform from the z-domain to the time domain, the inverse z-transform, two methods may be considered:

(a) Inverse transform by means of the inverse z-transform and the calculation rules.
(b) Inverse transform by means of division.

We shall explain these methods by two examples. There are no general rules that indicate which of the two solution methods works fastest in a certain case. Often it is a matter of personal preference.

(a) In inverse transform by means of the z-transform a trick is used, which may be phrased as follows: First 'keep in hand a factor z' in the nominator; then apply partial fraction to the 'rest' and transform, by means of the table, the expression obtained to the time domain.

Example 11.11
Given:

$$X(z) = \frac{z}{(z-1)(z-2)} \quad (11.45)$$

what is $[x(nT)]$?

Solution:
'Keep z in hand', so

$$X(z) = z \left[\frac{1}{(z-1)(z-2)} \right]$$

Apply partial fraction

$$X(z) = z \left[\frac{-1}{z-1} + \frac{1}{z-2} \right] = \frac{-z}{z-1} + \frac{z}{z-2}$$

By means of the transform table (rules 1 and 7) we find:

$$\sum_{n=0}^{\infty} x(nT) = \sum_{n=0}^{\infty} [-1 + 2^n] \quad (11.46)$$

or

$$[x(nT)] = 0, 1, 3, 7, 15, \ldots .$$

(b) Inverse transform by means of division is self-evident. This method will be explained by means of the data of the previous example.

Example 11.12
Given:

$$X(z) = \frac{z}{(z-1)(z-2)} \tag{11.47}$$

what is $[x(nT)]$

Solution:

$$X(z) = \frac{z}{(z-1)(z-2)} = \frac{z}{z^2 - 3z + 2}$$

Division:

```
z² − 3z + 2  |  z                         | z⁻¹ + 3z⁻² + 7z⁻³ + 15z⁻⁴ + ...
                z − 3 + 2z⁻¹
                ─────────────
                    3 − 2z⁻¹
                    3 − 9z⁻¹ + 6z⁻²
                    ───────────────
                         7z⁻¹ − 6z⁻²
                         7z⁻¹ − 21z⁻² + 14
                         ─────────────────
                               15z⁻² − 14
                                    etc.
```

So:

$$X(z) = z^{-1} + 3z^{-2} + 7z^{-3} + 15z^{-4} + \ldots$$

According to formula (11.34) we find the result also found in (11.46).

$$[x(nT)] = 0, 1, 3, 7, 15, \ldots \tag{11.48}$$

Notes:
1. The advantage of the first method for inverse transform from the z-domain to the t-domain is that, from the obtained expression for $x(nT)$, it is easy to determine a signal magnitude at a random sample time. By means of the division procedure this may become a difficult matter. As we shall see later, the first method cannot always be applied.
2. By the inverse z-transform, only information regarding the magnitudes of the relative signal at the sampling moments will arise. There will be no information on the signal between sampling moments.

11.8 PROBLEMS

1. Let us assume the signal $x(t)$ of Fig. 11.10. The duration of the sampling period is $\frac{1}{2}$ sec.; sampling begins at $t = 0$. Write $x^*(t)$ as an impulse series.

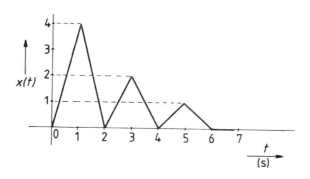

Fig. 11.10 — The signal $x(t)$ of problems 1 and 2.

2. Determine the Laplace transformed notation for $X^*(s)$ of the signal in Fig. 11.10.
3. We have the signal: $x(t) = 4 \cos 10t \, 1(t)$; the sampling frequency is 100 rad/s; sampling starts at $t = 0$.

   Determine the expressions $X^*(s)$ and $X^*(j\omega)$ for the sampled signal, and sketch the pole zero plot for $X^*(s)$.
4. Explain, by means of frequency spectra, Shannon's sampling theorem.
5. Determine, by means of expression (11.33), the z-transforms of:

   a $x(t) = 10 \, 1(t-2)$, with $T = 1$ s
   b $x(nT) = 2^n$
   c $x(t) = [x(nT)] = 0, 1, 2, 3, 2, 1, 0, 0 \ldots$
6. Determine, by means of the z-transform table, the z-transforms of the sampled form of:

   a $x(t) = 5 \, 1(t) + 2 \, 1(t - 2T)$
   b $x(t) = 5t(1 - e^{-2t}) \, 1(t)$
   c $x(t) = (4 \sin 5t + 8 \cos 5t) \, 1(t)$
   d $x(t) = 2 \, e^{-t} (\cos 3\pi t + \frac{1}{2} \sin 3\pi t) \, 1(t)$
   e $x(t) = t \sin \pi t \, 1(t)$
7. Determine, by means of the limit value theorems, $y(0)$ and $y(\infty)$, if

$$Y(z) = \frac{2z(z - e^{-T} \cos \pi T)}{z^2 - 2z \, e^{-T} \cos \pi T + e^{-2T}}$$

Sec. 11.8]   Problems   219

8. Determine, by means of inverse z-transform, by Table 11.1, signals $x(nT)$ of which the z-transforms are as follows:

a $\quad X(z) = \dfrac{2z}{(z-2)(z-4)}$.

b $\quad X(z) = \dfrac{z(z-1+2T)}{(z-1)^2}$.

c $\quad X(z) = \dfrac{2z(z - e^{-T} \cos \pi T + 2e^{-T} \sin \pi T)}{z^2 - 2z\, e^{-T} \cos \pi T + e^{-2T}}$

9. Determine by division, from the z-transforms below, the beginning of the series of $[x(nT)]$:

a $\quad X(z) = \dfrac{z^4 + 2z^3 + 3z^2 + 4z + 5}{z^5}$

b $\quad X(z) = \dfrac{z-1}{(z-2)(z-3)}$

c $\quad X(z) = \dfrac{z}{z^3 + 1}$

10. Determine the z-transform of $X^*(s)$, when:

$$X(s) = \dfrac{1 - e^{-sT}}{s} \cdot \dfrac{s+a}{s^2} .$$

11. We have a signal $f(t)$ with Laplace transformed transfer function:

$$F(s) = \dfrac{s+3}{(s+1)(s+2)}$$

Determine the transfer function $F(z)$ of the sampled signal, if the sampling period is $T = \ln 2$ seconds.

12. We have a signal $f(t) = t$. Determine by formula

$$F(z) = \sum_{n=0}^{\infty} f(nT) z^{-n}$$

the z-transformed transfer function of $f^*(t)$.

Note: $F(z)$ appears to consist of the product of two identical geometric series.

13. We have a sampled signal with z-transformed transfer function:

$$C(z) = \frac{z}{4(z - \tfrac{1}{2})^2 (z - 1)}$$

Question:
a Determine $c(0)$ up to and included $c(4T)$ both by division and by the table.
b Determine $c(\infty)$ by means of the final value theorem.

14. Repeat Exercise 13, but now for:

$$C(z) = \frac{-z^2 + 2z}{z^2 - 2z + 1}$$

12

Block diagrams of sampled data systems

12.1 INTRODUCTION

There are systems that operate with discrete signals only, but more often we meet with systems in which both discrete and continuous signals are found. The latter occur when signals are deliberately sampled in a system originally operating with continuous signals only.

In control systems based on digital process control, the part of the control loop between the measured signal (controlled variable) and the output signal of the digital controller operates with discrete signals with digital coded signal values.

As seen from the block diagrams, there are three relevant possibilities:

1. Every element of the block diagram is discrete, that is, equipped to process discrete signals. In this case all signals between the elements will be discrete.
2. Every element is continuous, that is, equipped to process continuous signals, but in one or more places signal sampling occurs.
3. There are both discrete and continuous elements, and discrete as well as continuous signals.

In the first case the description of the operation of every block in the block diagram may be done by means of a function of z, or by a transfer function $H(z)$. For the latter we apply the z-transform to series of numbers.

In the other two cases a problem occurs. For in continuous elements the input and output signals must be known at all times, whereas these signals are known only at the time of sampling in discrete elements. To analyse systems successfully in these two cases, we must consider only the magnitudes of these continuous signals at the time of sampling. Therefore we shall install a fictional sampling device in places where we are interested in the progress of an otherwise continuous signal, and then apply these values to our assessment. As before, we assume all sampling devices to be synchronous.

As we shall see in certain cases, in which we find both discrete signals and continuous elements, it will not be possible to determine the continuous output signal $y(t)$ at the time of sampling; thus $y(nT)$. To get some information on the progress of $y(t)$ between the times of sampling, the so called modified z-transform may be applied. However, we will not discuss this any further.

12.2 TRANSFER FUNCTIONS OF DISCRETE ELEMENTS AND SYSTEMS

Because in a discrete element the input and output signals are known only at the time of sampling, the z-transform may be applied directly to the input and output signals in order to determine $H(z)$, see Fig. 12.1.

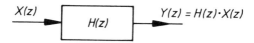

Fig. 12.1 — Block diagram of a discrete element in the z-domain.

The relation between the input and output is an algebraic equation (difference equation), and the following applies:

$$Y(z) = H(z) \cdot X(z) \tag{12.1}$$

and of course:

$$H(z) = \frac{Y(z)}{X(z)} \tag{12.2}$$

Example 12.1
Assume the input $x(nT)$ of a discrete element is given by $[x(nT)] = 2,1,0,0,0,\ldots$ and the output $y(nT)$ by: $[y(nT)] = 16,16,8,4,1,0,0,0,0,\ldots$.
What is the transfer function $H(z)$ of this element?

Answer:

$$X(z) = \mathscr{L}\{[x(nT)]\} = 2 + z^{-1}$$
$$Y(z) = \mathscr{L}\{[y(nT)]\} = 16 + 16z^{-1} + 8z^{-2} + 4z^{-3} + z^{-4}$$

By means of formula (12.2) follows:

Sec. 12.3] Transfer functions of sampled data systems

$$H(z) = \frac{Y(z)}{X(z)} = \frac{16 + 16z^{-1} + 8z^{-2} + 4z^{-3} + z^{-4}}{2 + z^{-1}} =$$

$$= \frac{16z^4 + 16z^3 + 8z^2 + 4z + 1}{2z^4 + z^3}$$

$$= 8 + 4z^{-1} + 2z^{-2} + x^{-3}$$

Example 12.2

Given the block diagram of Fig. 12.2, what is the transfer function $H(z) = \frac{C(z)}{R(z)}$.

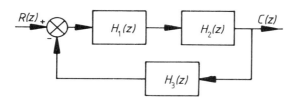

Fig. 12.2 — Block diagram of a feedback system with discrete elements in the z-domain.

Answer:
Every element in this block diagram is discrete, thus the relation according to formula (12.2) becomes:

$$H(z) = \frac{C(z)}{R(z)} = \frac{H_1(z) H_2(z)}{1 + H_1(z) H_2(z) H_3(z)}$$

12.3 TRANSFER FUNCTIONS OF SAMPLED DATA SYSTEMS

In a (continuous) element, being part of a sampled data-system, sampling may take place in three ways:

a Sampling at the output of the element.
b Sampling at the input of the element.
c Sampling at both input and output of the element.

(a) In Fig. 12.3 sampling at the output of a continuous element is shown.

Fig. 12.3 — Sampling at the output of a continuous element.

In Fig. 12.3:

$$Y(s) = H(s) \cdot X(s)$$

Then for the sampled output $Y^*(s)$:

$$Y(s) = \frac{1}{T} \sum_{n=-\infty}^{\infty} Y(s+jn\omega_s) = \frac{1}{T} \sum_{n=-\infty}^{\infty} H(s+jn\omega_s)X(s+jn\omega_s)$$

$$= [H(s) \cdot X(s)]^* \qquad (12.3)$$

After z-transform:

$$Y(z) = \mathscr{L}\{[H(s) \cdot X(s)]^*\} = HX(z) \qquad (12.4)$$

The form $HX(z)$ in expression (12.4) indicates that the z-transform of the product $H(s) \cdot X(s)$ must be determined. Note that this is not equal to $H(z) \cdot X(z)$. From formula (12.4) it appears that the transfer function of the element with transfer function $H(z)$ followed by sampling cannot be determined.

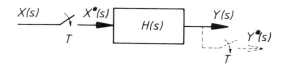

Fig. 12.4 — Sampling at the input of a continuous element.

Transfer functions of sampled data systems

(b) In Fig. 12.4, sampling at the input of a continuous element is shown thus:

$$Y(s) = H(s) \cdot X^*(s) \tag{12.5}$$

For this mixed expression it is not possible to carry out an inverse transform to the time domain; in other words, it is not possible to determine $y(t)$ in a simple fashion.

By means of the fictional sampling device mentioned in the introduction, it is quite possible to determine $Y^*(s)$. Then also the sampling values of $Y(t)$ at the time of sampling may be determined. Now:

$$Y^*(s) = \frac{1}{T} \sum_{n=-\infty}^{\infty} Y(s + jn\omega_s)$$

$$= \frac{1}{T} \sum_{n=-\infty}^{\infty} H(s + jn\omega_s) \cdot X^*(s + jn\omega_s) \tag{12.6}$$

By definition:

$$X^*(s + jn\omega_s) = X^*(s) \tag{12.7}$$

As this means sampling a sample, (12.6) becomes:

$$Y^*(s) = \frac{1}{T} \sum_{n=-\infty}^{\infty} H(s + jn\omega_s) \cdot X^*(s) = H^*(s) \cdot X^*(s) \tag{12.8}$$

Transfer to the z-domain yields:

$$Y(z) = H(z) \cdot X(z) \tag{12.9}$$

and also $\quad H(z) = \dfrac{Y(z)}{X)z)}$

(c) In Fig. 12.5 sampling at both input and output of a continuous element is shown.

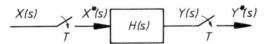

Fig. 12.5 — Sampling at both input and output of a continuous element.

Now:

$$Y(s) = H(s) \cdot X^*(s)$$

In view of the previous derivation of $Y^*(s)$, with fictional sampling device, the following also applies:

$$Y^*(s) = [H(s) \cdot X^*(s)]^* = H^*(s) \cdot X^*(s) \qquad (12.10)$$

and after z-transform:

$$H(z) = \frac{Y(z)}{X(z)} \qquad (12.11)$$

Again, $y(t)$ cannot be determined, but determination of $y(nT)$ is quite possible.
Up till now we have considered systems consisting of one block. Such a system will usually contain more blocks, and sampling may be done in various places in such a block diagram. In principle there are three different situations in block diagrams: blocks in cascade, parallel blocks, and feedback.

Blocks in cascade:
When two (continuous) blocks are in cascade it is important to know whether there is any sampling between these blocks (elements) or not. These two different situations are shown in Fig. 12.6. To determine $Y(z)$ and possibly $y(nT)$ we shall again use fictional sampling devices on the output signal.

Sec. 12.3] Transfer functions of sampled data systems

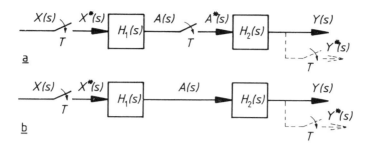

Fig. 12.6 — Blocks in cascade.

For Fig. 12.6a:

$$A(s) = H_1(s)X^*(s)$$

and $A^*(s) = H_1^*(s)X^*(s)$

$$Y(s) = H_2(s)A^*(s) = H_2(s)H_1^*(s)X^*(s)$$

and $Y^*(s) = H_2^*(s)H_1^*(s)X^*(s)$

After z-transform:

$$Y(z) = H_1(z)H_2(z)X(z)$$

or $\dfrac{Y(z)}{X(z)} = H_1(z) \cdot H_2(z)$ \hfill (12.12)

If, in a cascade of two blocks, sampling occurs between those blocks, the complete transfer function $H(z)$ equals the product of the separate transfer functions $H_1(z)$ and $H_2(z)$ of the blocks with transfer functions $H_1(s)$ and $H_2(s)$.

For Fig. 126b.:

$$\left. \begin{array}{l} A(s) = H_1(s)X^*(s) \\ \\ Y(s) = H_2(s)A(s) \end{array} \right\} \Rightarrow Y(s) = H_1(s)H_2(s)X^*(s)$$

So: $Y^*(s) = [H_1(s)H_2(s)]^* \cdot X^*(s)$

And after z-transform:

$$Y(z) = H_1H_2(z)X(z)$$

or $\dfrac{Y(z)}{X(z)} = H_1H_2(z)$

If in a cascade of blocks there is no sampling between the blocks, the complete transfer function $H(z)$ equals the z-transform of the complete transfer function $H(s) = H_1(s) \cdot H_2(s)$ of the two blocks with transfer functions $H_1(s)$ and $H_2(s)$.

Example 12.3
Given the block diagram as in Fig. 12.7;

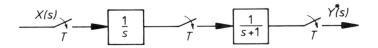

Fig. 12.7 — Block diagram for Example 12.3.

$$H(z) = \mathscr{L}\left\{\frac{1}{s}\right\} \cdot \mathscr{L}\left\{\frac{1}{z-1}\right\} = \frac{z}{z-1} \cdot \frac{z}{z-e^{-T}}$$

$$= \frac{z^2}{(z-1)(z-e^{-T})} \qquad (12.14)$$

Example 12.4
Given the block diagram of Fig. 12.8:

Fig. 12.8 — Block diagram of Example 12.4.

$$H(z) = \mathscr{L}\left\{\frac{1}{s(s+1)}\right\} = \mathscr{L}\left\{\frac{1}{s} - \frac{1}{s+1}\right\}$$

$$= \frac{z}{z-1} - \frac{z}{z-e^{-T}} = \frac{z(1-e^{-T})}{(z-1)(z-e^{-T})} \qquad (12.15)$$

Parallel blocks:
In Fig. 12.9 the parallel connection of two sampled (continuous) elements is shown. In this block diagram:

$$Y(s) = H_1(s)X^*(s) + H_2(s)X^*(s) = [H_1(s) + H_2(s)] \cdot X^*(s)$$

For the sampled output, $Y^*(s)$ applies:

Sec. 12.3] Transfer functions of sampled data systems

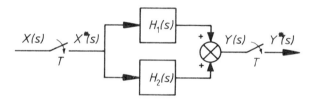

Fig. 12.9 — Parallel connection of two sample blocks.

$$Y^*(s) = [H_1(s) + H_2(s)]^* X^*(s)$$

And after z-transform:

$$Y(z) = \mathscr{Z}[H_1(s) + H_2(s)]^* \cdot X(z) = [H_1(z) + H_2(z)] \cdot X(z)$$

or: $\dfrac{Y(z)}{X(z)} = H_1(z) + H_2(z)$ (12.16)

So in parallel blocks as in Fig. 12.9, the z-transformed transfer function is the sum total of the separately z-transformed transfer functions $H_1(z)$ and $H_2(z)$.

Example 12.5

Given the block diagram of Fig. 12.9, with $H(s) = \dfrac{1}{s}$ and $H_2(s) = \dfrac{1}{s+1}$, the transfer function $H(z) = Y(z)/X(z)$ applies.

$$H(z) = \frac{z(2z - e^{-T} - 1)}{(z-1)(z - e^{-T})} \tag{12.17}$$

Feedback
With feedback in a sampled data system it is important to know whether sampling occurs in the feedback path or in the forward path. The former is shown in Fig. 12.10.

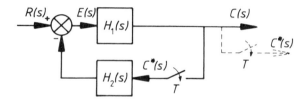

Fig. 12.10 — System with sampling feedback.

Here: $E(s) = R(s) - H_2(s)C^*(s)$ (12.18)
and: $C(s) = H_1(s)E(s($ (12.19)

From (12.18) and (12.19) follows:

$$C(s) = H_1(s)[R(s) - H_2(s)C^*(s)]$$
$$= H_1(s)R(s) - H_1(s)H_2(s)C^*(s)$$

For the sampled output:

$$C^*(s) = [H_1(s)R(s)]^* - [H_1(s)H_2(s)]^*C^*(s)$$

And after z-transform:

$$C(z) = H_1R(z) - H_1H_2(z)C(z)$$

or: $C(z)[1 + H_1H_2(z)] = H_1R(z)$

and: $$C(z) = \frac{H_1R(z)}{1 + H_1H_2(z)}$$ (12.20)

So in this case the transform function $H(z) = C(z)/R(z)$ cannot be determined because $R(z)$ does not occur separately in the numerator of the expression for $C(z)$.

Sampling in the forward path is represented in Fig. 12.11.

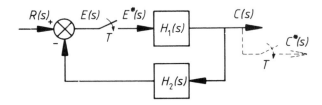

Fig. 12.11 — System with sampling in the forward path.

For the block diagram of Fig. 12.11:

$C(s) = H_1(s)E^*(s)$ (12.21)
and $E(s) = R(s) - H_2(s)C(s)$ (12.22)

From (12.21) and (12.22) follows:

Transfer functions of sampled data systems

so
$$E(s) = R(s) - H_2(s)H_1(s)E^*(s)$$
$$E^*(s) = R^*(s) - [H_1(s)H_2(s)]^*E^*(s)$$

and
$$E^*(s) = \frac{R^*(s)}{1 + [H_1(s)H_2(s)]^*}$$

For the sampled output, $C^*(s)$ applies:

$$C^*(s) = [H_1(s)E^*(s)]^* = H_1^*(s)E^*(s) \tag{12.23}$$

From (12.23) follows:

$$C^*(s) = \frac{H_1^*(s)}{1 + [H_1(s)H_2(s)]^*} \cdot R^*(s)$$

and after z-transform:

$$C(z) = \frac{H_1(z)}{1 + H_1H_2(z)} \cdot R(z)$$

So
$$H(z) = \frac{C(z)}{R(z)} = \frac{H_1(z)}{1 + H_1H_2(z)} \tag{12.24}$$

In the case of a unit feedback we have $H_2(s) = 1$, and

$$H(z) = \frac{C(z)}{R(z)} = \frac{H_1(z)}{1 + H_1(z)} \tag{12.25}$$

If in a block diagram signals and/or blocks are separated by sampling devices, the z-transforms will also occur separately in the transfer function; if they are not separated, z-transforms of the combination of non-separated signals and blocks occur.

Example 12.6

Let us assume that in the block diagram of Fig. 12.11 $H_1(s) = \frac{1}{s}$ and $H_2(s) = \frac{1}{s+1}$.

Then we find for $H(z)$:

$$H(z) = \frac{\mathscr{L}\left\{\frac{1}{s}\right\}}{1 + \mathscr{L}\left\{\frac{1}{s(s+1)}\right\}} = \frac{z(z - e^{-T})}{z^2 - 2z\,e^{-T} + e^{-T}} \qquad (12.26)$$

12.4 APPLICATIONS

The conclusions drawn in section 12.3 will now be applied to some examples.

Example 12.7

A frequent case in control engineering is shown in Fig. 12.12. Here a sampled signal

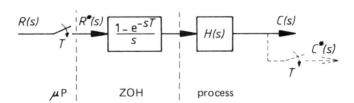

Fig. 12.12 — Controlling a process by means of a discrete control signal by a ZOH.

is fed to a zero order hold device (ZOH) followed by a process. Such a situation occurs in controlling a process by means of a microprocessor. At certain moments this microprocessor issues new control signals to the ZOH, the latter then issues a continuous control signal to the process.

Here we can calculate only the values of the process output signal at the time of sampling. Here:

$$\frac{C(z)}{R(z)} = \mathscr{L}\left[G_{ho}(s)H(s)\right] = \mathscr{L}\left[\frac{1 - e^{-sT}}{s} H(s)\right] \qquad (12.27)$$

Using the calculation rules of the z-transform, this leads to:

$$\frac{C(z)}{R(z)} = \mathscr{L}\left[\frac{H(s)}{s} - \frac{e^{-sT} H(s)}{s}\right] = \mathscr{L}\left[\frac{H(s)}{s}\right] - \mathscr{L}\left[\frac{e^{-sT} H(s)}{s}\right]$$

or: $$\frac{C(z)}{R(z)} = \mathscr{L}\left[\frac{H(s)}{s}\right] - z^{-1}\mathscr{L}\left[\frac{H(s)}{s}\right] = (1-z^{-1})\mathscr{L}\left[\frac{H(s)}{s}\right]$$

So: $$\frac{C(z)}{R(z)} = \frac{z-1}{z}\mathscr{L}\left[\frac{H(s)}{s}\right] \tag{12.28}$$

If, for example, the process employs integration, $H(s) = 1/s$ and becomes

$$\frac{C(z)}{R(z)} = \frac{z-1}{z} \cdot \mathscr{L}\left\{\frac{1}{s^2}\right\} = \frac{T}{z-1} \tag{12.29}$$

If, moreover, the input signal is a unit step, or: $r(t) = 1(t)$, then $R(s) = 1/s$ and becomes $R(z) = z/(z-1)$.

Then we have:

$$C(z) = \frac{T}{z-1} \cdot R(z) = \frac{T}{z-1} \cdot \frac{z}{z-1} = \frac{zT}{(z-1)^2} \tag{12.30}$$

And for the output signal of the process:

$$c(nT) = nT \tag{12.31}$$

which represents an easily verifiable, and in this case predictable, result.

Example 12.8
Given the diagram of Fig. 12.12, but with a first-order process with transfer function $H(s) = \frac{1}{s+1}$, determine the unit step response if also: $T = \ln 2$ seconds.

Answer:

$$\frac{C(z)}{R(z)} = \mathscr{L}\left[G_{ho}(s)H(s)\right] = \mathscr{L}\left[\frac{1-e^{-sT}}{s} \cdot \frac{1}{s+1}\right] =$$

$$= \frac{z-1}{z}\mathscr{L}\left[\frac{1}{s(s+1)}\right] = \frac{z-1}{z}\mathscr{L}\left[\frac{1}{s} - \frac{1}{s+1}\right] =$$

$$= \frac{z-1}{z}\left[\frac{z}{z-1} - \frac{z}{z-e^{-T}}\right] \tag{12.32}$$

If $R(s) = 1/s$ then $R(z) = \dfrac{z}{z-1}$, so:

$$C(z) = \frac{z}{z-1} \cdot \frac{z-1}{z} \left[\frac{z}{z-1} - \frac{z}{z-e^{-T}} \right] =$$

$$= \frac{z}{z-1} - \frac{z}{z-e^{-T}} \quad (12.33)$$

With $T = \ln 2$ seconds we obtain:

$$C(z) = \frac{z}{z-1} - \frac{z}{z-\frac{1}{2}} \quad (12.34)$$

Inverse transform by means of the table yields:

$$c(nT) = 1 - (\tfrac{1}{2})^n \quad (12.35)$$

And so: $[c(nT)] = 0, \tfrac{1}{2}, \tfrac{3}{4}, \tfrac{7}{8}, \tfrac{15}{16}, \ldots \quad (12.36)$

In this case inverse transform by division is a little bit more difficult:

$$C(z) = \frac{z}{z-1} - \frac{z}{z-\frac{1}{2}} = \frac{z}{2z^2 - 3z + 1} \quad (12.37)$$

After division we have:

$$C(z) = \tfrac{1}{2}z^{-1} + \tfrac{3}{4}z^{-2} + \tfrac{7}{8}z^{-3} + \ldots \quad (12.38)$$

which yields the same series $[c(nT)]$
The step response is shown in Fig. 12.13.

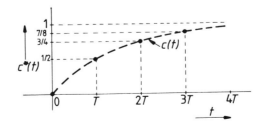

Fig. 12.13 — Step response of the system of Example 12.8.

Sec. 12.4] Applications 235

By the final value theorem the final value of $c(nT)$ becomes:

$$\lim_{n \to \infty} c(nT) = \lim_{z \to 1} \frac{z-1}{z} C(z) = \lim_{z \to 1} \frac{(z-1)}{z} \left[\frac{z}{z-1} - \frac{z}{z-\frac{1}{2}} \right]$$

$$= \lim_{z \to 1} \left[1 - \frac{z-1}{z-\frac{1}{2}} \right] = 1 \tag{12.39}$$

Of course this result also follows from the calculated expression for $c(nT)$ and a qualitative consideration of the considered system. Although we can say nothing about the progress of $c(t)$ between the sampling instants, in this simple case it is easy to see that $c(t)$ progresses along the bold dashed line in Fig. 12.13.

Example 12.9
Fig. 12.14 gives the block diagram of a sampled data system. The process consists of a

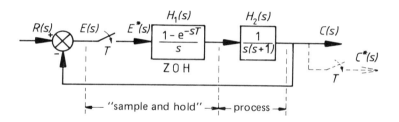

Fig. 12.14 — Sample data servosystem.

cascade of first order system in cascade ($K = 1$; $\tau = 1$ s) and a pure integrator. The feedback must see to it that the output follows the input as closely as possible (servosystem). The error signal $e = r - c$, however, is known only at discrete instants $(0, T, 2T, 3T, \ldots)$, in which the sampling period is $T = 1$ second.

A zero order hold device must reconstruct the signal, to provide a suitable control of the process.

Determine the response $c^*(t)$, if there is a step change in input $r(t)$ (unit step).

Answer
For this system (see section 12.3):

$$C(z) = \frac{H_1 H_2(z)}{1 + H_1 H_2(z)} \cdot R(z) \tag{12.40}$$

To determine $C(z)$ it is necessary to determine $H_1 H_2(z)$:

$$H_1H_2(z) = \mathscr{L}\left[\frac{1-e^{-sT}}{s} \cdot \frac{1}{s(s+1)}\right]$$

$$= (1-z^{-1})\mathscr{L}\left[\frac{1}{s^2(s+1)}\right]$$

$$= (1-z^{-1})\mathscr{L}\left[\frac{1}{s^2} - \frac{1}{s} + \frac{1}{s+1}\right]$$

$$= \left(\frac{z-1}{z}\right)\left[\frac{zT}{(z-1)^2} - \frac{z}{z-1} + \frac{z}{z-e^{-T}}\right]$$

When $T = 1$ we have:

$$H_1H_2(z) = \frac{z-1}{z}\left[\frac{z}{(z-1)^2} - \frac{z}{z-1} + \frac{z}{z-e^{-1}}\right]$$

$$= \frac{e^{-1}z + 1 - 2e^{-1}}{(z-1)(z-e^{-1})}$$

$$= \frac{0.37z + 0.26}{(z-1)(z-0.37)} \tag{12.41}$$

For $R(z)$ ($R(s) = 1/s$):

$$R(z) = \frac{z}{z-1} \tag{12.42}$$

Substitution of (12.41) and (12.42) into (12.40) yields:

$$C(z) = \frac{0.37z + 0.26}{(z-1)(z-0.37) + 0.37z + 0.26} \cdot \frac{z}{z-1}$$

$$= \frac{0.37z^2 + 0.26z}{z^3 - 2z^2 + 1.63z - 0.63}$$

After division we have:

$$C(z) = 0.37z^{-1} + z^{-2} + 1.4z^{-3} + 1.4z^{-4} + 1.15z^{-5} + 0.9z^{-6} + 0.8z^{-7} + \ldots \tag{12.43}$$

Using the final value theorem for $c(\infty)$ yields:

$$c(\infty) = \lim_{z \to 1} \frac{z-1}{z} C(z) = \lim_{z \to 1} \frac{0.37z + 0.26}{(z-1)(z-0.37) + 0.37z + 0.26} = 1 \quad (12.44)$$

This fits, because the process contains a pure integrator, so that the feedback system has a dc gain of 1.
So for $[c(nT)]$:

$$[c(nT)] = 0;\ 0.37;\ 1;\ 1.4;\ 1.4;\ 1.15;\ 0.9;\ 0.8;\ \ldots \quad (12.45)$$

In Fig. 12.15 we see two step responses, obtained by computer simulation, of the

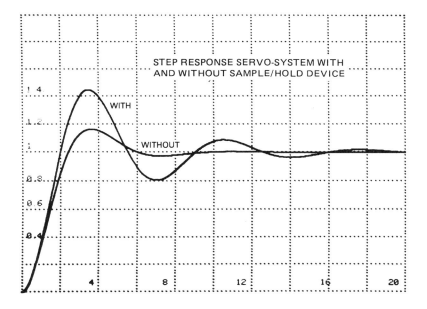

Fig. 12.15 — Step responses of a computer simulation of the servosystem with and without sample/hold device.

servosystem discussed before. The overshoot in the step response is about 45% with sample and hold, and about 16% without. It is clear that the system with sample and hold is less damped.

Fig. 12.16 shows the complete response of the control signal after the sample and hold device. The step-wise behaviour of this reconstructed signal is clear.

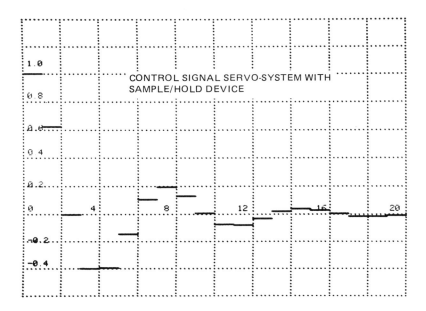

Fig. 12.16 — The reconstructed control signal of the servosystem.

12.5 PROBLEMS

1. Assume the input signal of a discrete element is: $x(nT) = 1, 0, 1, 0, 0, \ldots$ and the matching output signal is: $y(nT) = 0, 0, 2, 0, 2, 0, 0, \ldots$. Determine the transfer function $H(z)$ of this element.
2. Given: the block diagram of Fig. 12.17; the input is a unit step; the sampling

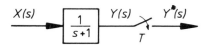

Fig. 12.17 — Block diagram of Problem 2.

period is $T = 1$ second.
   Determine the output at the sampling instants: $y(nT)$. Also determine $y(\infty)$. Was this a predictable result?
3. Given: the block diagram of Fig. 12.18; the sampling period is $T = 1$ second.

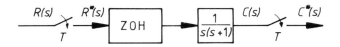

Fig. 12.18 — Block diagram of Problem 3.

Determine $H(z) = C(z)/R(z)$.
4. Given: the block diagram of Fig. 12.19. $T = \ln 2$ seconds.

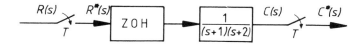

Fig. 12.19 — Block diagram of Problem 4.

Determine the unit step response $[c(nT)]$ and also $c(\infty)$.
5. Given: the block diagram of a sampled data system, according to Fig. 12.20. The sampling

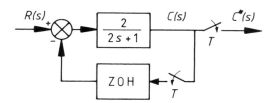

Fig. 12.20 — Block diagram of problem 5.

period is $T = 1$ second.
a Determine $C(z)$, if there is a unit step on the input.
b Determine $c(nT)$ for $n = 0, 1, 2, 3$.
6. Given: the block diagram of Fig. 12.21; the sampling period is $T = 0.5$ second.

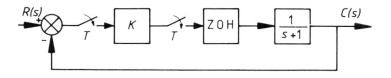

Fig. 12.21 — Block diagram of Problem 6.

a Determine the transfer function $H(z) = C(z)/R(z)$.
b Determine $c(nT)$, if there is a unit step on the input, for $K = 1$, 1.5, and 2.5.
c Determine for those values of K also $c(\infty)$, and sketch the responses.

13
Analysis and synthesis of sampled data systems in the z-domain

13.1 INTRODUCTION

The methods applied for the analysis and synthesis of sampled data systems are principally the same as those in continuous systems. We shall see that for a system to be stable there will be requirements as to the position of the poles of this system in the z-domain. To design sampled data systems we adopt an approach similar to that applied for continuous ones. The design criteria may be translated from the t-domain or s-domain to the z-domain. By means of the root locus theory now applied in the z-domain we may then design the necessary control actions.

13.2 STABILITY ANALYSIS IN THE z-DOMAIN

Earlier we defined the stability of a system as follows:

A linear time invariant system is stable if a finite input signal results in a finite output signal.

To examine the stability of sampled data systems we can use direct and indirect methods. In a direct method the stability is directly appreciated by means of the complete transfer function of the system, just as was done in continuous systems (e.g. Routh's criterium). We shall use one indirect method, with which the stability is appreciated by the pole/zero representation in the z-plane. By applying the root locus method we may also determine the loop gain at which the system tends to become unstable.

First we must deciede what requirement must be met by the pole/zero representation for a system to be stable. For this purpose we use the z-transformed transfer function $H(z)$ of the system.

Because in linear systems the stability is independent of the character of the input

Sec. 13.2]   Stability analysis in the z-domain

signal, the stability investigation may also take place in a function not determined explicitly. Then appreciation is done by means of the z-transform of the output signal at a constant input signal, for then we find:

$$C(z) = RH(z) = \mathscr{L}\{R(s) \cdot H(s)\} = R \cdot \mathscr{L}\{Hs\} = R \cdot H(z) \tag{13.1}$$

The transfer function of a system is represented with $m \leq n$ by:

$$H(z) = \frac{C(z)}{R(z)} = \frac{b_m z^m + b_{m-1} z^{m-1} + \ldots + b_1 z + b_0}{a_n z^n + a_{n-1} z^{n-1} + \ldots + a_1 z + a_0} \tag{13.2}$$

and so:

$$C(z) = H(z) \cdot R(z) \tag{13.3}$$

If the input $R(z)$ is finite, this may be written as a fraction with finite numerator and denominator polynomial, in which the highest degree of z in the numerator is at most equal to the degree (r) of z in the denominator.

If we assume the denominator polynomial of $H(z)$ to have single roots p_i (poles of the systems), after partial fraction we have:

$$C(z) = \frac{A_1 z}{z - p_1} + \frac{A_2 z}{z - p_2} + \ldots = \sum_{i=1}^{n+r} \frac{A_i z}{z - p_i} \tag{13.4}$$

with: n = number of poles of $H(z)$
r = number of poles of input signal $R(z)$
p_i = poles of the output signal.

For coinciding poles in $H(z)$ the above expression does not apply. This exclusion does not affect the following argument. Taking into account coinciding roots leads to more extensive and less surveyable derivation. For the output at the time of sampling after inverse z-transform:

$$c(kT) = A_1 p_1^k + A_2 p_2^k + A_3 p_3^k + \ldots = \sum_{i=1}^{n+r} A_i p_i^k \tag{13.5}$$

In a stable system the output is finite, so that in expression (13.5) must apply:

$$|p_i| < 1 \quad \text{for} \quad i = 1, 2, 3, \ldots, n+r \tag{13.6}$$

In that case the expression for $C(z)$, according to (13.4), represents a converging geometric series. So the moduli of the poles p_i must be smaller than one. From inequality (13.6) now follows the important conclusion as to the stability of a sampled data system:

A linear time invariant sampled data system is stable if all the poles of that system lie within the unit circle in the z-plane.

The stability requirement corresponds with the known requirement as to the stability of a linear time invariant continuous system, namely that all poles of that system must be in the left half-plane of the s-plane.
In Fig. 13.1 this symmetry of both stability criteria is shown. There must be no poles in the shaded areas.

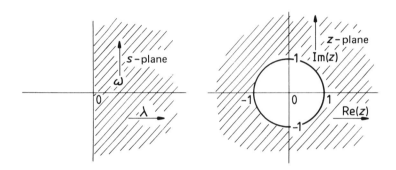

Fig. 13.1 — Comparison of the forbidden areas for poles of a stable continuous system and a stable sampled data system.

As the reader may check for himself, the same stability follows from the consideration that for a discrete system the poles of $H^*(s)$ should all be in the left half-plane, whereas this whole part of the s-plane is represented by the z-transform formula (11.26) in the unit circle in the z-plane.

A rough estimation of the stability of a sampled data system, with a ZOH for the reconstruction of the sampled signal, may be made by the approximation of sampling and reconstruction by an (extra) time delay of T (see also section 10.7). If the poles and zeros of the open loop system are known, it is possible to find out, by means of a root locus construction, when by any chance the closed loop system becomes unstable for certain values of the loop gain.

The construction rules for sketching a root locus of a continuous system (s-plane) also apply in the z-plane. As these construction rules originated from the root locus equation of the controlled system, it does not matter whether this equation is a polynomial in s or in z.

Example 13.1
Given the system of Fig. 13.2, we shall see how the stability of this system depends on

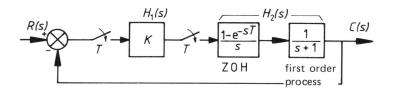

Fig. 13.2 — A proportionally controlled sampled data first order process.

the choice of sampling period T.
    For the transfer function:

$$H(z) = \frac{C(z)}{R(z)} = \frac{H_1(z) \cdot H_2(z)}{1 + H_1(z) \cdot H_2(z)} \tag{13.7}$$

with

$$H_1(z) = K \tag{13.8}$$

$$H_1(z) = \mathscr{L}\left\{\frac{1-e^{-sT}}{s} \cdot \frac{1}{s+1}\right\}$$

$$= (1-z^{-1})\mathscr{L}\left\{\frac{1}{s(s+1)}\right\}$$

$$= (1-z^{-1})\left\{\frac{1}{s} - \frac{1}{s+1}\right\}$$

$$= \frac{z-1}{z}\left\{\frac{z}{z-1} - \frac{z}{z-e^{-T}}\right\}$$

$$= \frac{1-e^{-T}}{z-e^{-T}} \tag{13.9}$$

Substitution of (13.8) and (13.9) into (13.7) yields:

$$H(z) = \frac{K\dfrac{1-e^{-T}}{z-e^{-T}}}{1 + K\dfrac{1-e^{-T}}{z-e^{-T}}} = \frac{K(1-e^{-T})}{z - e^{-T}(1+K) + K} \tag{13.10}$$

The root locus equation for variation of K becomes:

$$z - e^{-T}(1 + K) + K = 0 \qquad (13.11)$$

or

$$\frac{1 - e^{-T}}{z - e^{-T}} = -\frac{1}{K} \qquad (13.12)$$

There is one pole in $z = +e^{-T}$; this is the starting point of the one root locus section which the root locus consists of. The root locus for variation of K from 0 to ∞ is shown in Fig. 13.3; the unit circle is sketched also.

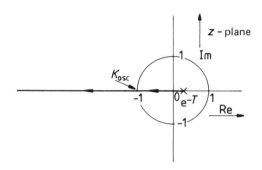

Fig. 13.3 — Root locus for variation of K from 0 to ∞ of a proportionally controlled first order process.

For $T = 1$ the starting point lies at the pole: $z = e^{-1} \simeq 0.368$. The system becomes unstable for this value of T if (see expression (13.11)):

$$z = e^{-1}(1 + K) - K = -1 \qquad (13.13)$$

from which follows

$$K_{osc} \simeq 2.2 \qquad (13.14)$$

For the relation between sampling period T and the gain factor at which the system will oscillate, the following generally applies:

$$e^{-T}(1 + K_{osc}) - K_{osc} = -1$$

or also

Sec. 13.2] Stability analysis in the z-domain

$$K_{osc} = \frac{1+e^{-T}}{1-e^{-T}}$$

This relation is shown graphically in Fig. 13.4. As was to be expected, the stability will worsen as the sampling period becomes longer.

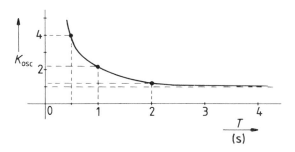

Fig. 13.4 — Relation between K_{osc} and T.

Example 13.2
The sampled data second order system of Fig. 13.5 is given. For this case we shall also

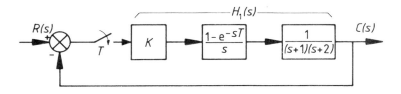

Fig. 13.5 — A proportionally controlled sampled data second order system.

determine for which value of K the system becomes unstable when the sampling period is $T = \ln 2$ s.

For the transfer function $H(z) = \dfrac{C(z)}{R(z)}$:

$$H(z) = \frac{H_1(z)}{1+H_1(z)} \tag{13.15}$$

with

$$H_1(z) = \mathscr{L}\left\{\frac{1-e^{-sT}}{s} \cdot \frac{K}{(s+1)(s+2)}\right\}$$

$$= K(1-z^{-1})\mathscr{L}\left\{\frac{1}{s(s+1)(s+2)}\right\}$$

$$= K(1-z^{-1})\mathscr{L}\left\{\frac{\frac{1}{2}}{s} - \frac{1}{s+1} + \frac{\frac{1}{2}}{s+2}\right\}$$

$$= K(1-z^{-1})\mathscr{L}\left\{\frac{\frac{1}{2}z}{z-1} - \frac{z}{z-e^{-T}} + \frac{\frac{1}{2}z}{z-e^{-2T}}\right\} \qquad (13.16)$$

For $T = \ln 2$ s:

$$H_1(z) = K\left(\frac{z-1}{z}\right)\left(\frac{\frac{1}{2}z}{z-1} - \frac{z}{z-\frac{1}{2}} + \frac{\frac{1}{2}z}{z-\frac{1}{4}}\right)$$

$$= \frac{K(z+\frac{1}{2})}{8(z-\frac{1}{2})(z-\frac{1}{4})} \qquad (13.17)$$

Thus the root locus for variation of K from 0 to ∞ becomes:

$$\frac{(z+\frac{1}{2})}{8(z-\frac{1}{2})(z-\frac{1}{4})} = -\frac{1}{K} \qquad (13.18)$$

In Fig. 13.6 the root locus in the z-plane is sketched together with the unit circle.

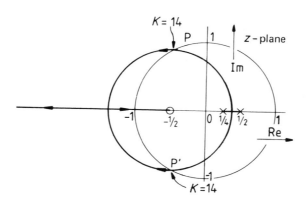

Fig. 13.6 — Root locus of the sampled data second order system for $0 < K < \infty$.

From this figure we may understand that the system becomes unstable if the root locus intersects the unit circle at points P and P'. Then there are two complex conjugate poles $z_{1,2} = a \pm jb$. From the root locus equation (13.18) it follows:

$$8(z-\tfrac{1}{2})(z-\tfrac{1}{4}) + K(z+\tfrac{1}{2}) = 0 \tag{13.19}$$

and $\qquad 16z^2 + (2K-12)z + 2 + K = 0 \tag{13.20}$

The complex conjugate poles $z_{1,2}$ are obtained from equation (13.20):

$$z_{1,2} = \frac{-(2K-12) \pm \sqrt{(2K-12)^2 - 64(2+K)}}{32} =$$

$$= \frac{-(2K-12) \pm j\sqrt{64(2+K) - (2K-12)^2}}{32} = a \pm jb \tag{13.21}$$

The modulus of these poles is: $|z_{1,2}| = \sqrt{a^2 + b^2}$. At the points P and P' this modulus is 1, so

$$\frac{1}{32}[(2K-12)^2 + 64(2+K) - (2K-12)^2]^{\frac{1}{2}} = \tfrac{1}{4}\sqrt{2+K} \equiv 1 \tag{13.22}$$

from which follows: $K = 14$. Then the poles are:

$$z_{1,2} = -\tfrac{1}{2}(1 \pm j\sqrt{3}) \tag{13.23}$$

From the root locus of Fig. 13.6 it appears that the system is unstable for any value of $K > 14$.

13.3 DESIGN CRITERIA IN THE z-DOMAIN

It is obvious that we should apply the criteria of absolute and relative damping, defined in the s-domain, in the z-domain too. For this we first consider how a line of absolute damping in the s-domain is represented in the z-domain. For such a line:

$$s = \lambda_1 + j\omega \tag{13.24}$$

with constant λ_1 and variable ω. According to the z-transform:

$$z = e^{sT} \tag{13.25}$$

and from (13.24) and (13.25):

$$z = e^{(\lambda_1 + j\omega)T} \tag{13.26}$$

For the modulus of z:

$$|z| = e^{\lambda_1 T} \tag{13.27}$$

For the argument of z:

$$\arg z = \omega T.$$

So, if a line of absolute damping is completed in the s-domain, a full circle with its centre in the origin will be completed in the z-domain. For $\lambda_1 = 0$ this circle is a unit circle; for every value of λ_1 for which applies $\lambda_1 < 0$ a circle that lies within this unit circle is completed. If in the s-domain a line of absolute damping is completed from $-\frac{1}{2}\omega_s$ to $+\frac{1}{2}\omega_s$, this is just the part of the considered line of absolute damping that lies in the primary strip in the s-domain. Then the argument runs in the z-domain from:

$$-\tfrac{1}{2}\omega_s T = -\pi \quad \text{to} \quad +\tfrac{1}{2}\omega_s T = +\pi \tag{13.28}$$

Moreover:

$$e^{sT} = e^{sT + j2\pi n} = e^{sT + jn\omega_s T} \tag{13.29}$$

Every part of a line of absolute damping in a secondary strip is represented on the same circle in the z-domain as the part in the primary strip. If $\lambda_1 = 0$ the whole imaginary axis in the s-domain is represented, by z-transform, on the unit circle in the z-domain, and the whole left half-plane is represented within this unit circle.

As for the radius of the circle with absolute damping λ_1:

$$R_c = e^{\lambda_1 T} \tag{13.30}$$

and for $\lambda_1 < 0$ this radius will get smaller as the absolute damping increases.

So if a certain minimum absolute damping λ_1 is required, all poles of the controlled system in the z-domain must lie within the circle with radius $R_c = e^{\lambda_1 T}$, and its centre in the origin.

From (13.30) directly follows that, for a required specific absolute damping λ_1 and increasing value of the sampling period, the area in which the poles may lie gets smaller. We may also say, that for a fixed value of the scampling period the absolute damping will be larger as the poles lie within a smaller circle with its centre in the origin.

Example 13.3
Given the block diagram of Fig. 13.7a, with the sampling period of 0.1 s. For which value of K will this system have two complex conjugate poles that lie on the circle with radius $R_c = \tfrac{7}{8}$ and what will be the value of t_s (2%)?

Sec. 13.3] Design criteria in the z-domain 249

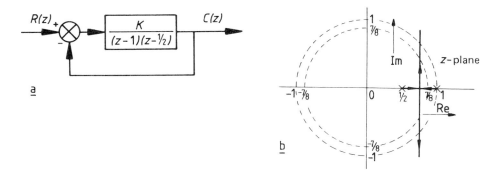

Fig. 13.7 — (a) Block diagram of Example 13.3. (b) Root locus of Example 13.3.

Answer
From the root locus of the controlled system (13.7b) it follows that for the poles:

$$z_{1,2} = \tfrac{3}{4} \pm j\sqrt{(\tfrac{7}{8})^2 - (\tfrac{3}{4})^2}$$

or $\quad z_{1,2} = 0.75 \pm j0.45$ \hfill (13.31)

The root locus equation is:

$$(z-1)(z-\tfrac{1}{2}) + K = 0 \hfill (13.32)$$

Equalization of the root locus equation of (13.32) and the one from (13.31) yields:

$$z^2 - 1.5z + 0.5 + K = 0$$
$$z^2 - 1.5z + (0.75)^2 + 0.45^2 = 0$$

or $\quad 0.5 + K = 0.5625 + 0.2025$

So: $\quad K = 0.265$ \hfill (13.33)

For λT we find from:

$$e^{\lambda T} = 0.875, \quad \text{so} \quad \lambda T = -0.1335 \hfill (13.34)$$

As $T = 0.1$ second:

$$\lambda = -1.335 \hfill (13.35)$$

For $t_s(2\%)$:

$$e^{\lambda t_s(2\%)} = \frac{1}{50} \quad \text{or also} \quad \lambda t_s(2\%) \simeq -4$$

from which follows:

$$t_s(2\%) \simeq 3 \text{ seconds} \tag{13.36}$$

For a line of constant relative damping in the s-domain:

$$\frac{\lambda}{\omega} = \text{constant} \tag{13.37}$$

By means of the relative damping the overshoot in a step response may be determined. For now:

$$D = e^{\frac{\lambda}{\omega}\pi} \cdot 100\% \tag{13.38}$$

Now let us see how a line of constant relative damping is represented in the z-domain. Here:

$$z = e^{sT} = e^{(\lambda+j\omega)T} = e^{\lambda T} \cdot e^{j\omega T} \tag{13.39}$$

So: $\quad |z| = e^{\lambda T} \tag{13.40}$

and $\quad \arg z = \omega T \tag{13.41}$

For varying values of $\lambda/\omega = q$, logarithmic spirals that lie within the above unit circle will develop in the z-plane. Some of these spirals are given in Fig. 13.8

It might be possible to denote a set of curves of constant relative damping in the z-plane by parameters q and T. However, this yields no easily applicable criterion. So mostly we shall confine ourselves to a method in which we shall design for a certain desired settling time, and afterwards we shall define what the relative damping and the matching overshoot in the step response will be like. If the overshoot is too large, a larger absolute damping will be chosen, and a new check of the overshoot follows.

Example 13.4
Determine the overshoot in the step response for the setting of K in Example 13.3

Solution:
For both poles, when $K = 0.265$:

Sec. 13.4] Proportional control design of sampled data first order system

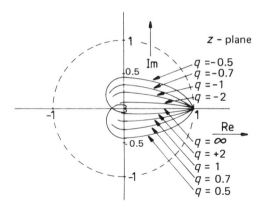

Fig. 13.8 — Lines of constant relative damping in the z-plane.

$$z_{1,2} = 0.75 \pm j0.45$$

According to (13.41) we obtain:

$$|z| = e^{\lambda T} = \sqrt{0.75^2 + 0.45^2} = 0.875$$

or: $\lambda T = -0.13$ (13.42)

Furthermore we have:

$$\arg(z) = \arctan \frac{0.45}{0.75} = 0.54 \text{ rad} = \omega T$$

so $\dfrac{\lambda}{\omega} = \dfrac{\lambda T}{\omega T} = -\dfrac{0.13}{0.54} = -0.24$ (13.43)

and the overshoot becomes:

$$D = e^{\frac{\lambda}{\omega}\pi} \cdot 100 \simeq 47\%.$$

13.4 PROPORTIONAL CONTROL DESIGN OF A SAMPLED DATA FIRST ORDER SYSTEM

Let us assume a digitally controlled first order system with a time constant of 1 s and a dc gain of 1, as shown in Fig. 13.9a. The set point is put into the computer numerically as the measured process value is converted by means of an ADC preceded by a sampling device. Reconstruction of the control signal of the process calculated in the CPU is obtained by a ZOH. The latter is followed by a DAC. Every

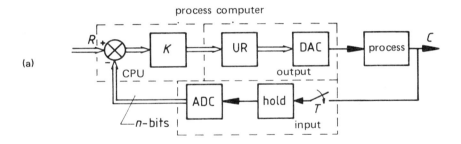

(a)

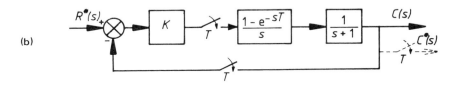

(b)

Fig. 13.9 — (a) Digitally controlled first order process. (b) Block diagram of a digitally controlled first order process.

T seconds the hold device (a digital register) is provided with a newly calculated control value; T is set at 0.5 s.

The block diagram of the controlled system is shown in Fig. 13.9b. As calculated before (13.11), in Example 13.1, for the root locus we have:

$$\frac{1-e^{-T}}{z-e^{-T}} = -\frac{1}{K}$$

and for $T = 0.5$ s

$$\frac{0.393}{z - 0.607} = -\frac{1}{K} \tag{13.44}$$

The root locus for variation of K from zero to infinity is shown in Fig. 13.10, and also the unit circle (stability).

Assume that such setting of the proportional gain is required that the overshoot in the response is maximally 20%; now calculate the settling time $t_s(2\%)$.

To determine the overshoot we shall use expressions (13.40) and (13.41).

$$|z| = e^{\lambda T} \quad \text{and} \quad \arg(z) = \omega T$$

For an overshoot of 20%:

$$D = 20\% = e^{\frac{\lambda}{\omega}\pi} \cdot 100\%,$$

Sec. 13.4] **Proportional control design of sampled data first order system**

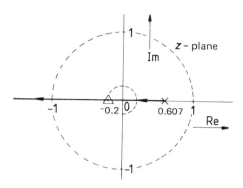

Fig. 13.10 — Root locus of a P controlled sampled data first order process.

from which follows:

$$\frac{\lambda}{\omega} \simeq -0.512 \qquad (13.45)$$

So long as the pole lies on the positive real axis in the z-plane there is no overshoot, so in any case the gain factor may be increased to such an extent that the pole will be on the negative real axis. Then we have:

$$\arg(z) = \pi = \omega T \qquad (13.46)$$

With the value for λ/ω found in (13.45) we obtain:

$$\lambda T = \frac{\lambda}{\omega} \cdot \pi \simeq -1.61 \qquad (13.47)$$

and

$$|z| = e^{-1.61} \simeq 0.2 \text{ (see also Fig. 13.10)}$$

So the pole may be at $z = -0.2$; substituting this result into the root locus equation (13.44) yields:

$$K \simeq 2 \qquad (13.48)$$

With $T = 0.5$ s we obtain from (13.47):

$$\lambda \simeq -3.22$$

For $t_s(2\%)$ we have:

$$e^{\lambda t_s(2\%)} = \frac{1}{50}, \quad \text{or:}$$

$$t_s(2\%) \simeq 1.2 \text{ s} \tag{13.49}$$

To check the calculated setting we shall calculate the step response of the controlled system with $K=2$.

For the transfer function of the controlled system applies, see expression (13.10):

$$H(z) = \frac{0.8}{z + 0.2} \tag{13.50}$$

Now the unit step response becomes:

$$C(z) = \frac{0.8z}{(z-1)(z+0.2)} \tag{13.51}$$

By partial fraction and inverse transform or by division, $c(nT)$ may be determined; we find:

$$
\begin{aligned}
&t = 0 \;:\; c(0) = 0 & &t = 1.5 \;:\; c(1.5) = 0.66 \\
&t = 0.5 \;:\; c(0.5) = 0.8 & &t = 2.0 \;:\; c(2.0) = 0.67 \\
&t = 1.0 \;:\; c(1.0) = 0.64 & &t = 2.5 \;:\; c(2.0) = 0.67 \quad \text{etc.}
\end{aligned}
$$

In Fig. 13.11 the step responses for some values of K, including $K=2$, have been plotted. It is clear that the overshoot and the settling time for $K=2$ correspond well with the calculated values. A special case occurs when $K=1.5$; then the pole of the feedback system will be at $z=0$. The transfer function of the system will be:

$$H(z) = \frac{0.6}{z} = 0.6z^{-1} \tag{13.52}$$

This represents a time delay of T seconds and a gain factor of 0.6. The step response again represents a step with respect to the magnitudes at the sampling instants; now, however, shifted in time by a period T. The output $c(nT)$ follows the input, if this changes by steps, after exactly one sampling period. This response is known as deadbeat response; see Fig. 13.11 ($K=1.5$); this will be discussed in Chapter 16.

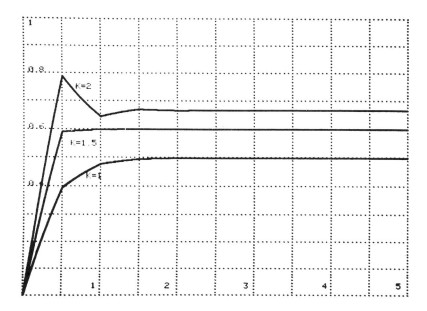

Fig. 13.11 — Some step responses of the proportionally controlled sampled data first order process.

13.5 PROPORTIONAL CONTROL DESIGN OF A SAMPLED DATA SECOND ORDER SYSTEM

In Fig. 13.12 we have the block diagram of a controlled sampled data second order process. It is essentially the same system of which the stability aspect has been considered in Example 13.2.

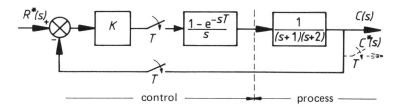

Fig. 13.12 — Block diagram of the proportionally controlled sampled data second order process.

We assume the sampling period to be $T = \ln 2 \text{ s}$, and such value for the proportional gain factor (K) that the settling time (5%) will be 3 s, whereas the overshoot is calculated.

According to (13.18) the root locus equation of the controlled system, for variation of K from 0 to ∞ is:

$$\frac{(z+0.5)}{8(z-0.5)(z-0.25)} = -\frac{1}{K}$$

The root locus, already shown in Fig. 13.6, is sketched again in Fig. 13.13.

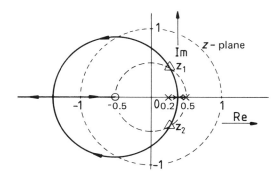

Fig. 13.13 — Root locus of a proportionally controlled sampled data second order process.

The design requirement: $t_s(5\%) = 3$ s results in the design criterion for the z-plane that the pole must lie on the circle with its centre in the origin and a radius of:

$$R_c = e^{\lambda_1 T} \tag{13.53}$$

in which for λ_1 applies:

$$e^{\lambda_1 t_s(5\%)} = \frac{1}{20}, \quad \text{or} \quad \lambda_1 \simeq -1 \tag{13.54}$$

With $T = \ln 2$ s and $\lambda_1 = -1$ we have:

$$R_c = 0.5 \tag{13.55}$$

Thus the poles may be at the intersections z_1 and z_2 of the root locus and the circle with radius $R = 0.5$ and its centre in the origin (see Fig. 13.13). In this case the position of the poles may easily be measured from the figure because the root locus can be sketched exactly. The circle part has its centre in $z = -0.5$ and the radius is $\sqrt{0.75} = 0.86$. We find:

Sec. 13.5]   **Design of a sampled data second order system**   257

$$z_{1,2} \simeq 0.25 \pm j0.43 \tag{13.56}$$

From the root locus equation we find:

$$K = 2 \tag{13.57}$$

Please check the results of (13.56) and (13.57)!
We calculate the expected overshoot for this choice of K; here applies:

$$|z_{1,2}| = e^{\lambda T} = 0.5$$

and $\quad \arg(z_{1,2}) = \arctan \dfrac{\pm 0.43}{0.25} = \pm 1.04 \text{ rad} \tag{13.58}$

From (13.58) follows:

$$\lambda T = -0.693$$

and $\quad q = \dfrac{\lambda}{\omega} = \dfrac{\lambda T}{\omega T} = \dfrac{-0.693}{1.04} \simeq -0.67$

and the overshoot becomes:

$$D = e^{\frac{\lambda}{\omega}\pi} \cdot 100\% \simeq 12\% \tag{13.59}$$

The controlled system has, at the calculated setting ($K=2$), the following transfer function:

$$H(z) = \dfrac{z + \frac{1}{2}}{4(z^2 - \frac{1}{2}z + \frac{1}{4})} \tag{13.60}$$

If we put a unit step with $R(z) = \dfrac{z}{z-1}$ on the input the response becomes:

$$C(z) = \dfrac{z(z + \frac{1}{2})}{4(z-1)(z^2 - \frac{1}{2}z + \frac{1}{4})} \tag{13.61}$$

By division we find the following sampling values:

$$c(nT) = 0; 0.25; 0.5; 0.56; 0.53; 0.5; \ldots$$

The step response is shown in Fig. 13.14; this step response confirms the correctness of the setting and the satisfaction of the design criteria.

The design of integral and derivative control actions is essentially the same in both the z-domain and the s-domain. In the z-domain too the path of root

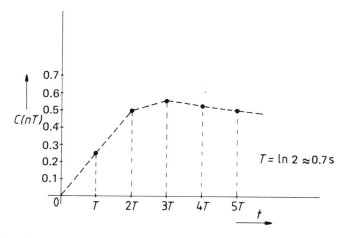

Fig. 13.14 — Step response of the P-controlled sampled data second order system for $K = 2$.

locus sections may be influenced by adding pole/zero combinations in such a way that more favourable conditions of the design criteria are satisfied.

In many cases we shall change over to a more practical approach (see Chapter 15).

13.6 PROBLEMS

1. Given the block diagram of a sampled data system as in Fig. 13.15, with:

$$D(z) = \frac{2(z-1)}{2z^2 + z} \qquad (13.62)$$

a. Sketch the pole/zero plot of the open loop system.
b. Determine the transfer function $H(z) = \dfrac{C(z)}{R(z)}$.

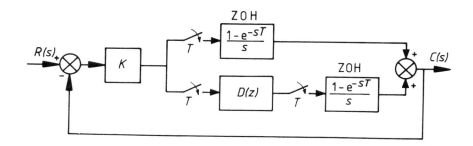

Fig. 13.15 — Block diagram of the sampled data system of Problem 1.

c. Investigate for which positive value of K the system will be on the edge of (in)stability.

2. Given the block diagram of a sampled data system as in Fig. 13.16. The sampling period is $T = 3 \ln 2$ s.
   a. Determine K if the controlled system has a pole at $z = \tfrac{1}{4}$.
   b. For which positive values of K will the system be stable. Give reasons for your answer.

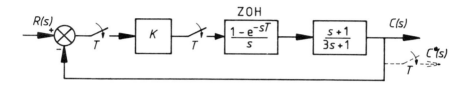

Fig. 13.16 — Block diagram of the sampled data system of Problem 2.

3. Given the block diagram of a sampled data system as in Fig. 13.17; the sampling period T is: $T = 0.5$ s.

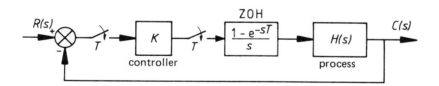

Fig. 13.17 — Block diagram of the sampled data system of Problem 3.

The transfer function of the process is:

$$H(s) = \frac{e^{-sT}}{1 + \tau s}, \quad \text{with} \quad \tau = 2.24 \text{ s} \tag{13.63}$$

a. Determine the z-transformed transfer function of the open loop system
b. Determine the positive values of K for which the system is unstable.
c. Calculate the value of K so that the settling time $t_s(5\%) = 5$ sec.
d. Determine the overshoot in the step response found for K in (c).

e. Assume the system to be set so that no overshoot occurs; then determine the value of K and calculate t_s (5%).

4. Given the block diagram of a sampled data system as in Fig. 13.18, with $T = 2 \ln 2$ s.

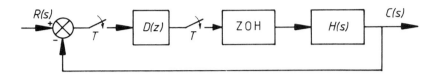

Fig. 13.18 — Block diagram of the sampled data system of Problem 4.

Furthermore, we have:

$$D(z) = \frac{2(3-z^{-1})}{3(1-z^{-1})} \quad \text{and} \quad H(s) = \frac{K}{2s+1} \tag{13.64}$$

a. Determine for $K = 1.5$ the transfer function $H(z) = \dfrac{C(z)}{R(z)}$.
b. Sketch the root locus of this system for K from $0 \to \infty$.
c. For which positive values of K will the system be unstable?

5. Given the block diagram of a proportionally controlled first order system with time delay as in Fig. 13.19, with $T = 1$ s.

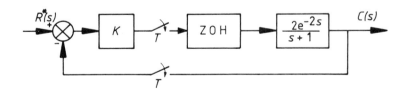

Fig. 13.19 — Block diagram of the sampled data system of Problem 5.

a. Determine the transfer function $H(z) = \dfrac{C(z)}{R(z)}$.
b. Sketch the root locus of this system for variation of K from 0 to ∞.
c. For which positive values of K will this system be unstable?

14
Digital networks

14.1 INTRODUCTION

The most important application of a computer in a control loop is the realization of digital processing of signals in a direct digital control (DDC). Such a realization is often called a digital network. The functions of such a network are:

- Realization of the control action (PID) or a compensation network.
- Processing and filtering of signals.

In this chapter we shall mainly consider how digital networks are realized; the operation of such networks will mostly be shown in realization diagrams.

If digital processing is done by means of a digital computer certain aspects must be taken into account. In modern computer techniques the memory size (RAM) and the accuracy are hardly ever a problem. More important is the calculation speed. This depends on the type of computer and on the programming, as we shall see later.

Realization diagrams show us how calculations are done in the Central Processing Unit (CPU) of the computer. From a realization diagram an application program may be written.

As the realization of digital processing — depending on the computer's program — may be done in different ways, this is indicated only schematically. We shall not go into the programming itself. The main processes of digital realization are:

- Adding and subtracting of numbers.
- Holding a number of values during a number of sampling periods.
- Multiplying by a constant factor.

These processes may easily be realized in a computer.

Our starting point is the z-transformed transfer function $D(z)$ that must be realized.

$$D(z) = \frac{Y(z)}{X(z)} = \frac{b_m z^m + \ldots + b_1 z + b_0}{a_n z^n + \ldots + a_1 z + a_0} \tag{14.1}$$

in which $Y(z)$ is the output signal of the digital network and $X(z)$ the input signal. After division:

$$D(z) = \frac{Y(z)}{X(z)} = c_i z^{m-n} + c_{i-1} z^{m-n-1} + \ldots \text{ with } c_i = \frac{b_m}{a_n} \text{ etc.} \qquad (14.2)$$

If we consider the term z^{-k} to correspond with a time shift (delay) for k sampling periods, this means that $m \leq n$ in any case, because otherwise a positive power occurs in z, which would indicate a prediction. An important requirement for the transfer function $D(z)$ to be realized is that the highest degree of z in the numerator. If this requirement is satisfied, $D(z)$ may be realized by writing the relation between the input and output as a difference equation, and subsequently the output at the sampling instants, $y(nT)$ as a function of the other variables. This will be explained by means of an example.

Example 14.1
Given the transfer function:

$$D(z) = \frac{Y(z)}{X(z)} = \frac{az}{z+b} \qquad (14.3)$$

We may write $D(z)$ as:

$$D(z) = \frac{a}{1 + bz^{-1}}$$

or:

$$Y(z) + bz^{-1} Y(z) = aX(z) \qquad (14.4)$$

Inverse transformation to the time domain yields the following difference equation:

$$y(nT) + by(nT - T) = ax(nT)$$

or:

$$y(nT) = ax(nT) - by(nT - T) \qquad (14.5)$$

The difference equation of (14.5) indicates that the output $y(nT)$ equals a times the sampled input value $x(nT)$ minus b times the preceding output value $y(nT - T)$.

In Fig. 14.1 the realization diagram of expression (14.5) is shown. The multiplication factors a and b are shown in circles, the time delay of T seconds (a memory location!) in a square.

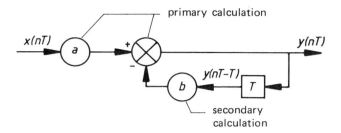

Fig. 14.1 — Realisation diagram of $D(z) = \dfrac{az}{z+b}$.

Although it does not matter very much how such a simple realization diagram (Fig. 4.1) is programmed, it is important to know that calculation by computer causes an extra time delay. Also, the attainability of the requirements for the sampling time T depends on it. To shorten the delay, caused by the computer calculations, as much as possible we must distinguish primary and secondary calculations. For Example 14.1 these are shown in Fig. 14.1. According to difference equation (14.5) the uppermost 'path' in this diagram is needed 'directly'. The new term $(nT-T)$ is needed only at the next sampling instant, so that this calculation may be done *between* sampling instants. The realization as in Fig. 14.1 is an example of direct programming. In the next section some other programming method will be discussed. The programming method depends on the application (required speed) and the type of (micro-) computer. So, for example, the number of available registers in the CPU may determine the choice of the programming method.

14.2 SOME PROGRAMMING METHODS

According to the realization of a digital network with transfer function $D(z)$ we distinguish:

- direct programming,
- serial programming,
- parallel programming,
- canonical programming.

We shall discuss these programming methods and illustrate them in an example.

Direct programming
The starting point in direct programming is the expression for the output signal at the sampling instants:

$$y(nT) = \sum_{k=0}^{m} a_k x(nT - kT) - \sum_{k=1}^{n} b_k y(nT - kT) \tag{14.6}$$

By means of this expression the momentary value of the output signal may be calculated from a number of sampling values of the input signal and a number of sampling values of the output signal.

Example 14.2
Let us consider the digital network with transfer function:

$$D(z) = \frac{Y(z)}{X(z)} = \frac{1 - 0.25z^{-1}}{(1 - z^{-1})(1 - 0.5z^{-1})} \tag{14.7}$$

From expression (14.7):

$$Y(z) - 1.5z^{-1} Y(z) + 0.5z^{-2} Y(z) = X(z) - 0.25z^{-1} X(z) \tag{14.8}$$

and after inverse transformation:

$$y(nT) = 1.5y(nT - T) - 0.5y(nT - 2T) + x(nT) - 0.25x(nT - T) \tag{14.9}$$

The realisation diagram is shown in Fig. 14.2.

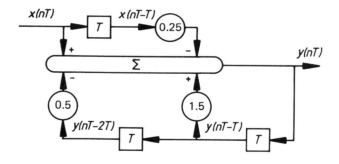

Fig. 14.2 — Realization diagram of $D(z)$ according to direct programming.

Serial programming
Our starting point for serial programming is splitting the transfer function $D(z)$ into factors $D_i(z)$ and $G_j(z)$, so that we have

$$D_i(z) = \frac{1}{d_i z^{-1} + 1} \text{ and } G_j(z) = c_j z^{-1} + 1 \quad \begin{matrix} i = 1, \ldots, n \\ j = 1, \ldots, m \end{matrix} \tag{14.10}$$

$D(z)$ is then:

Sec. 14.2]   Some programming methods   265

$$D(z) = \delta G_1(z) \cdot G_2(z) \ldots G_m(z) \cdot D_1(z) \cdot D_2(z) \ldots D_n(z) \quad (14.11)$$
with δ = constant

The transfer functions $D_i(z)$ and $G_j(z)$ are realized separately and then put in cascade.

Example 14.3
Assuming the transfer function $D(z)$ in Example 14.2, we have:

$$D(z) = \frac{1 - 0.25z^{-1}}{(1 - z^{-1})(1 - 0.5z^{-1})} = (1 - 0.25z^{-1}) \cdot \frac{1}{1 - z^{-1}} \cdot \frac{1}{1 - 0.5z^{-1}}$$

This is shown in Fig. 14.3.

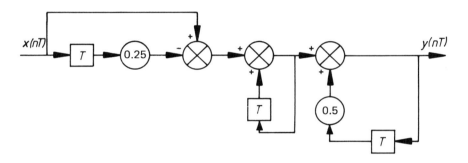

Fig. 14.3 — Realization diagram of $D(z)$ according to serial programming.

Parallel programming

In parallel programming transfer function $D(z)$ is split into the following form:

$$D(z) = \sum_{i=1}^{n} D_i(z) \quad (14.12)$$

in which $D_i(z)$ has the same form as in serial programming; the transfer functions $D_i(z)$ are now connected in parallel.

Example 14.4
The transfer function $D(z)$ has again of the same form as in Examples 14.2 and 14.3, and is written, by partial fraction, in the form:

$$D(z) = \frac{1.5}{1-z^{-1}} - \frac{0.5}{1-0.5z^{-1}}$$

The realization is shown in Fig. 14.4.

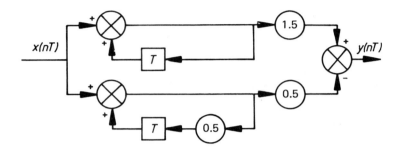

Fig. 14.4 — Realization diagram of $D(z)$ according to parallel programming.

Canonical programming
We start from the general notation of the transfer function $D(z)$:

$$D(z) = \frac{Y(z)}{X(z)} = \frac{b_0 + b_1 z^{-1} + \ldots + b_m z^{-m}}{a_0 + a_1 z^{-1} + \ldots + a_n z^{-n}} \qquad (14.13)$$

From this follows the difference equation:

$$a_0 y(kT) + a_1 y(kT-T) + \ldots + a_n y(kT-nT) = $$
$$= b_0 x(kT) + b_1 x(kT-T) + \ldots + b_m x(kT-mT) \qquad (14.14)$$

or:

$$y(kT) = \frac{b_0}{a_0} x(kT) + \left\{ \frac{b_1}{a_0} x(kT-T) - \frac{a_1}{a_0} y(kT-T) \right\} + \ldots$$
$$+ \left\{ \frac{b_2}{a_0} x(kT-2T) - \frac{a_2}{a_0} y(kT-2T) \right\} + \ldots \qquad (14.15)$$

The general realization of (14.15) is shown in Fig. 14.5

Example 14.5
Assuming the transfer function $D(z)$ realized before, we now find:

$$y(nT) = x(nT) + \{-0.25x(nT-T) + 1.5y(nT-T)\} - 0.5y(nT-2T)$$

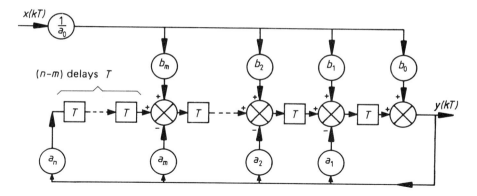

Fig. 14.5 — General realisation according to canonical programming.

The realization diagram is shown in Fig. 14.6.

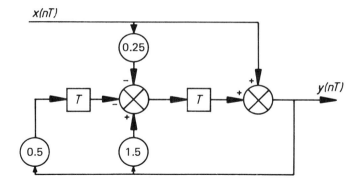

Fig. 14.6 — Realization diagram of $D(z)$ according to canonical programming.

14.3 CONTROL ALGORITHMS

With direct digital control (DDC) the control actions must be realized in the process computer. The necessary control algorithms may be derived from the realizations by continuous controllers. In principle a proportional control action remains a multiplication of the error signal by a constant factor (proportional gain). An integral control action develops, in principle, from determining the sum total of a number of sampling values of the error signal. A derivative control action can be realized by determination of the differences of successive samples of the error signal.

Except for the control actions in the control algorithm, the subtraction point is also realized in the computer by comparison of the set point and the measured value.

In practice we distinguish two groups of algorithms: position control algorithms and velocity control algorithms. In a position control algorithm the output signal of the algorithm is a direct measure of the position of the actuator, for example, a valve. If the process computer breaks down the actuator will go to zero position. In a velocity control algorithm the output signal of the algorithm is a measure of the adjustment of the actuator with regard to the previous position. If in this case the process computer breaks down, the actuator will stay in the previous position. Because the control algorithm provides only an adjustment signal, the actuator must be of such construction that the previous position will be held.

If the actuator is a voltage controlled element then, for example, a stepping motor with attached potentiometer may serve as receiver of the control signal from the process computer. The adjustment signal issued by the control algorithm must be converted into a number of pulses with which the stepping motor makes as many steps for the potentiometer to provide the correct control voltage for the actuator. If the actuator is a valve, this might also be driven by a stepping motor to realize a velocity algorithm. In practice the velocity algorithm appears to be the more usable. Its advantages over the position algorithm are:

1. At a break-down (failure, maintenance) of the digital controller the last data on the output signal will be saved, because only changes in the control signal are passed on. The last position of the actuator will be maintained.
2. Switching to an analog controller (for example as a back-up system) or manual control is done without sudden changes (nominal controller output equal to zero), so that possible damage to the actuator and/or saturation might occur in the process (this is a so called bumpless transfer).
3. The actuator (for example, stepping motor) itself provides the hold action, so that this may be controlled directly from the computer. (No DA converter and hold device are needed).

In Fig. 14.7 a,b two different configurations are shown.

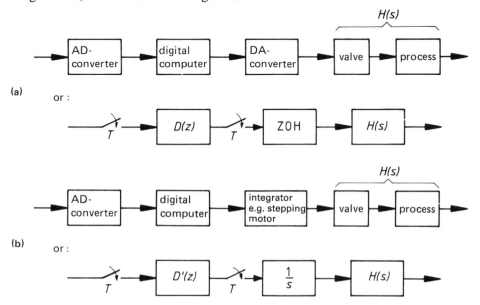

Fig. 14.7 — (a) Configuration and block diagram position control algorithm. (b) Configuration and block diagram velocity control algorithm.

Control algorithms

Before converting the above control algorithms into PID algorithms, we shall first consider the control actions separately. The input signal will always be represented by e and the output signal by y. Deriving the discrete control action is always done by discreting the corresponding analog control action. This discreting is only allowed if the sampling interval T is sufficiently short.

Separate control actions according to the position algorithm
For a proportional action (gain) the output of the controller must be directly proportional to the input, according to:

$$y(t) = k_p e(t), \text{ so: } y(nT) = k_p e(nT) \tag{14.16}$$

in which k_p = proportionality factor.
In the z-domain this yields:

$$Y(z) = k_p E(z) \text{ or } \frac{Y(z)}{E(z)} = k_p \tag{14.17}$$

In an integral action the error signal is integrated; this means addition in the discrete control action, so:

$$y(t) = \frac{1}{\tau_i} \int_0^t e(t)dt, \text{ so } y(nT) = \frac{T}{\tau_i} \sum_{k=0}^{n} e(kT) = k_1 \sum_{k=0}^{n} e(kT) \tag{14.18}$$

in which $k_i = \frac{T}{\tau_i}$ = integration factor.
Also

$$y(nT - T) = k_i \sum_{k=0}^{n-1} e(kT) \tag{14.19}$$

Subtraction of (14.18) and (14.19) yields:

$$y(nT) - y(nT - T) = k_i e(nT) \tag{14.20}$$

In the z domain this yields:

$$Y(z) - z^{-1}Y(z) = k_i E(z)$$

or:

$$\frac{Y(z)}{E(z)} = \frac{k_i}{1 - z^{-1}} \tag{14.21}$$

In a derivative action the error signal is differentiated and in the discrete case the output of the controller is determined as proportional to the difference between the controller input at $t = nT$ and at $t = nT - T$, so:

$$y(t) = \tau_d \frac{de(t)}{dt} \quad \text{and:} \quad y(nT) = \tau_d \frac{(e(nT) - e(nT - T))}{T} =$$

$$= k_d(e(nT) - e(nT - T)) \tag{14.22}$$

in which $k_d = \dfrac{\tau_d}{T} =$ differentiation constant.

In the z-domain this yields:

$$Y(z) = k_d(E(z) - z^{-1}E(z)) \quad \text{or} \quad \frac{Y(z)}{E(z)} = k_d(1 - z^{-1}) \tag{14.23}$$

Separate control actions according to the velocity control algorithm
Instead of the complete control signal for the position of the actuator we must now generate the change from the previous position with regard to the output signal.

The formulas for the velocity algorithm may simply be derived from those for the position algorithm. The controller output is now indicated by $y'(nT)$ and the transfer function of the controller by $D'(z)$.

The *P action* for the position algorithm becomes:

$$y(nT) = k_p e(nT) \tag{14.24}$$

and also

$$y(nT - T) = k_p e(nT - T) \tag{14.25}$$

so that:

$$y'(nT) = y(nT) - y(nT - T) = k_p(e(nT) - e(nT - T)) \tag{14.26}$$

In the z-domain this yields:

$$Y'(z) = k_p(E(z) - z^{-1}E(z)) \text{ or } D'(z) = \frac{Y'(z)}{E(z)} = k_p(1 - z^{-1}) \qquad (14.27)$$

The I action becomes from formula (14.20):

$$y'(nT) = y(nT) - y(nT - T) = k_i e(nT) \qquad (14.28)$$

In the z-domain this yields:

$$Y'(z) = k_i E(z) \text{ or: } D'(z) = \frac{Y'(z)}{E(z)} = k_i \qquad (14.29)$$

The D action becomes:

$$y(nT) = k_d(e(nT) - e(nT - T))$$

and

$$y(nT - T) = k_d(e(nT - T) - e(nT - 2T))$$

so that

$$y'(nT) = y(nT) - y(nT - T) = k_d(e(nT) - 2e(nT - T) + e(nT - 2T)) \qquad (14.30)$$

In the z-domain this yields:

$$Y'(z) = k_d(E(z) - 2z^{-1}E(z) + z^{-2}E(z))$$

or:

$$D'(z) = \frac{Y'(z)}{E(z)} = k_d(1 - 2z^{-1} + z^{-2}) \qquad (14.31)$$

In Table 14.1 the control actions are shown again.

From Table 14.1 it is clear how we may realize a discrete control algorithm, assuming desired proportional, integral, and derivative actions.

Comparison of the analog and digital control configuration, see Fig. 14.8, shows us that in fact the effect of the hold action must also be taken into account in designing the control action.

If the sampling frequency can be chosen sufficiently high, so that the influence of

Table 14.1 — Control actions for position and velocity control algorithms

$$k_i = \frac{T}{\tau_i} \quad k_d = \frac{\tau_d}{T}$$

control action	position algorithm	velocity algorithm
P: $y(nT) = k_p e(nT)$	k_p	$k_p(1 - z^{-1})$
I: $y(nT) = y(nT - T) + k_i e(nT)$	$k_i/1 - z^{-1}$	k_i
D: $yn(T) = k_d(e(nT) - e(nT - T))$	$k_d(1 - z^{-1})$	$k_d(1 - 2z^{-1} + z^{-2})$

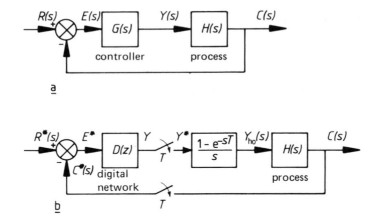

Fig. 14.8 — Comparison of analog and digital process control.

the (zero order) hold device may be neglected, the control actions may/can be designed by means of the results in Table 14.1.

The values of the integration time and the differentiation time are converted into factors k_i and k_d by multiplication and division by T respectively. The integration and differentiation times, have the same values as if the process were analog controlled.

Sec. 14.3] **Control algorithms** 273

If it is expected that, by sampling, such effects may arise, that these need, to be taken into account in designing the control action, we may act as follows:

1. The sampling effects are integrated as an extra time delay of $\frac{1}{2}T$, for calculations in the Bode diagram or in the adjustment rules (see Chapter 15). The values found for τ_i and τ_d will then be converted into k_i and k_d respectively, and we may continue by means of Table 14.1.
2. A design of the control actions in the z-domain follows, for which we are not actually committed to a conventional control action, but a (realizable) combination of zero(s) and poles may be chosen.

Notes:
1. The above approximations of control actions may easily be used for realizing digital signal processing. The continuous operation $\frac{df}{dt}$ is then replaced by the approximation:

$$\frac{f(nT) - f(nT - T)}{T} \tag{14.32}$$

Example 14.6
Let us realise the transfer function:

$$H(s) = \frac{s+1}{2s+1} = \frac{Y(s)}{X(s)} \tag{14.33}$$

Here applies: $sX(s) + X(s) = 2s\, Y(s) + Y(s)$
And the differential equation is: $\frac{dx(t)}{dt} + x(t) = 2\frac{dy(t)}{dt} + y(t)$
Discreting yields:

$$\frac{x(nT) - x(nT-T)}{T} + x(nT) = 2\frac{y(nT) - y(nT-T)}{T} + y(nT) \tag{14.34}$$

With desired output $y(nT)$ we find:

$$y(nT) = \frac{1}{T+2}\{2y(nT-T) + (T+1)x(nT) - x(nT-T)\} \tag{14.35}$$

After z-transformation follows:

$$D(z) = \frac{Y(z)}{X(z)} = \frac{1 + T - z^{-1}}{2 + T - 2z^{-1}} \tag{14.36}$$

2. If the function to be realised digitally is given in the form of a summation of derivatives of input and output signals, the z-form may be achieved even faster by the substitution:

$$s = \frac{1 - z^{-1}}{T} = \frac{z - 1}{zT}$$

This follows from applying the z-transform to expression (14.32).

Example 14.7
Realising the operation as desired in expression (14.33) then also gives:

$$D(z) = \frac{\frac{z-1}{zT} + 1}{2\frac{z-1}{zT} + 1} = \frac{z - 1 + zT}{2z - 2 + zT} = \frac{1 + T - z^{-1}}{2 + T - 2z^{-1}}$$

This is (of course) the same result as in (14.36).

14.4 EXAMPLES OF CONTROL ALGORITHMS

By means of the relations already derived, and summarized in Table 14.1, the digital control algorithms will be derived from the transfer functions of some continuous controllers.

Example 14.8
Let us assume the case of a continuous PI controller with a transfer function:

$$G(s) = K\left(1 + \frac{1}{s\tau_i}\right) = K + \frac{K}{s\tau_i} \qquad (14.37)$$

For the discrete realization as a position control algorithm we have, according to the tables

$$D(z) = k_p + \frac{k_p \cdot k_i}{1 - z^{-1}}, \text{ with } k_p = K \text{ and } k_i = \frac{T}{\tau_i} \qquad (14.38)$$

From which follows:

$$D(z) = \frac{a - bz^{-1}}{1 - z^{-1}} \qquad (14.39)$$

with: $a = k_p(1 + k_i)$

and: $\quad b = k_p$

If the velocity algorithm is used, then:

$$D'(z) = k_p (1 - z^{-1}) + k_p k_i$$

$$= k_p (1 + k_i) - k_p z^{-1}$$

$$= a - bz^{-1} \tag{14.40}$$

in which a and b have the same values as in (14.39).

Example 14.9
We assume a PID cascade controller with transfer function:

$$G(s) = K\left(1 + \frac{1}{s\tau_i}\right)(s\tau_d + 1)$$

$$= K\left(1 + \frac{\tau_d}{\tau_i} + s\tau_d + \frac{1}{s\tau_i}\right) \tag{14.41}$$

The discrete realization according to the position algorithm yields, by means of the table:

$$D(z) = k_p \left[k'_p + k_d(1 - z^{-1}) + \frac{k_i}{1 - z^{-1}} \right]$$

in which:

$$k_p = K$$

$$k'_p = 1 + \frac{\tau_d}{\tau_i} \tag{14.42}$$

$$k_d = \frac{\tau_d}{T}$$

and

$$k_i = \frac{T}{\tau_i}$$

From this follows: $k_d k_i = \tau_d/\tau_i$ and $k'_p = 1 + k_d k_i$

Then the expression for $D(z)$ becomes:

$$D(z) = \frac{Y(z)}{E(z)} = k_p \left[\frac{(1+k_d k_i)(1-z^{-1}) + k_i + k_d(1-z^{-1})^2}{1-z^{-1}} \right]$$

$$\frac{k_p(1 + k_d k_i + k_i + k_d) - k_p(2k_d + 1 + k_i k_d)z^{-1} + k_p k_d z^{-2}}{1-z^{-1}} \quad (14.43)$$

14.5 IMPLEMENTATION OF CONTROL ALGORITHMS

Regarding the practical application of digital control algorithms, we must take certain precautions to preclude undesired phenomena. These may be:

- The drifting of the process output by the application of a velocity algorithm if successive measuring values hardly differ. For example, if the output signal of a process is almost constant or if the sampling frequency is relatively high. If in such a case a P or PD control algorithm is assumed, the control signal may be so small that the process will not respond to it. This may occur if the resolution of the applied DAC is too small or because the actuator has a certain threshold value.

  A remedy for that phenomenon is the addition of an I action to the control algorithm. This will prevent drifting, as the reader may verify for himself.

- Digital oscillation. This may occur as a result of the finite resolution of the applied ADC and DAC. In a closed control loop oscillation will arise by which the LSB becomes 0 and 1 alternately. Of course an ADC and DAC with higher resolution decreases the amplitude of the oscillation. We might also apply converters with non-linear conversion. This is done because the conversion of low amplitude signals yields a relatively larger error than signals with higher amplitudes. The ADC is then preceded by a signal compressor. After (uniform) quantising in the ADC, the converted signal is expanded again.

- Output wind-up. This may arise in digital controllers because an overflow situation arises in the computer; for example, as a result of prolonged integration of a positive or negative error signal. Thereby the output signal may fall to zero. The remedy for this is, as in analog controllers, to apply signal limitation in the computer. Moreover the integrator will often need to be desaturated because otherwise it might take too long before it produces normal output values again, which might lead to a large overshoot in the process output.

  The effect of rounding off errors in the calculation by the control algorithm may be reduced by taking into account more bits than expressed in the output signal. Too small a sampling period will lead to such mistakes rather soon (see, for example, the small difference between coefficients a and b in the PI algorithm of (14.39) if $T \ll \tau_i$).

Furthermore, for practical implementation of the controller, we must also take into account a number of factors that increase the utility of the system:

- There must be a possibility to transfer the control loop from automatic control (AUT) to manual control (MAN) and vice versa. Moreover this transfer must be as 'bumpless' as possible.
- It must be possible to choose freely between a position and a velocity control algorithm, depending on the actuator which is activated by the control algorithm.
- The derivative control action must preferably have no influence if the error signal changes because of an adjustment of the desired value (setpoint change). This is achieved by integrating the derivative action in the feedback loop. The dynamic behaviour of the control system (loop transfer function) does not change because of this. This may also be done for the proportional control action. This is called 'avoiding proportional and derivative kick'.
- To speed up the desaturation of the integral action (see output wind-up) we may recalculate for this addition all the time, in case the maximum (MAX) or minimal (MIN) value of the control signal is exceeded. This may also be done during manual operation, so that bumpless transfer is achieved.
- From a practical point of view we mostly apply the parallel configuration of the PID controller. The summation of the proportional (PROR), the integral(INT), and derivative (DER) control signal for the actuator. It will be possible to set the P action, I action, and D actions separately. The actions do not interact. Please note that from a design point of view it is often more convenient to choose the cascade configuration, because this fits better into the pole/zero considerations. Both controllers may be 'transformed into each other', see also section 15.3.

In Fig. 14.9 we have a practical realization of a PID algorithm, in which most of the above aspects are incorporated.

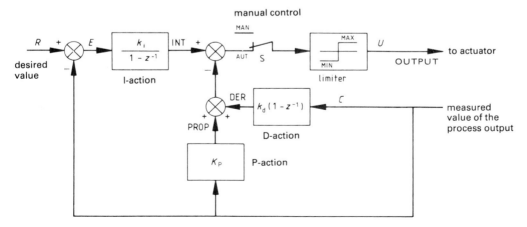

Fig. 14.9 — Practical realization of a PID algorithm.

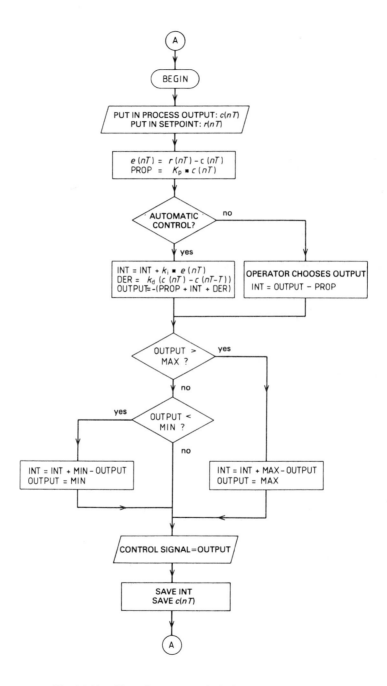

Fig. 14.10 — Flow diagram practical PID controller.

Problems

In Fig. 14.10 a flow diagram is shown, in which the above recalculations have been included. From this flow diagram the necessary programming may easily be written in the desired application software.

14.6 PROBLEMS

1a. Make a realization diagram of the following transfer function by direct programming

$$H(z) = \frac{z^2 + z - 2}{z^2 + 3} \tag{14.44}$$

b. Do the same for:

$$H(z) = \frac{1}{z^3} \tag{14.45}$$

c. And again the same for:

$$H(z) = \frac{z^2 - z + 2}{z - 1} \tag{14.46}$$

2. Given the realization diagram of a digital network as in Fig. 14.11.
    Determine the transfer function $Y(z)/X(z)$.
3. Given the following realization diagram of a digital network as in Fig. 14.12.
    a. Determine the transfer function $Y(z)/X(z)$.
    b. Draw the realization diagram according to canonical programming.
4. Given the circuit of a continuous control action as in Fig. 14.13.
    The operational amplifier is assumed to be 'ideal'.
    Determine the transfer function $D(z)$ of the digital control action with the same function as the given circuit for both the position and the velocity control algorithm. Assume the sampling period is $T = 0.25$ s.
5. Determine $D'(z) = Y(z)/E(z)$ for the digital velocity control algorithm corresponding with the continuous PID controller of:

$$y(t) = k_p \left\{ e(t) + \frac{1}{\tau_i} \int_0^t e(t) dt + \tau_d \frac{de(t)}{dt} \right\} \tag{14.47}$$

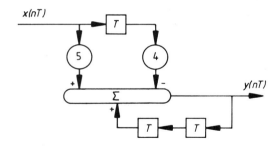

Fig. 14.11 — Realization diagram of the digital network of Problem 2.

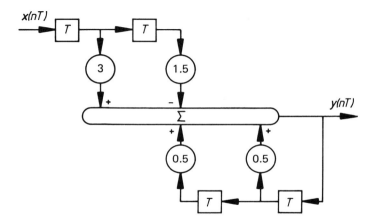

Fig. 14.12 — Realization diagram of the digital network of Problem 3.

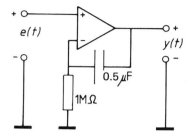

Fig. 14.13 — Electronic realization of a continuous control action.

6. Determine $D(z) = Y(z)/E(z)$ for the digital position algorithm corresponding to the continuous PD controller with the transfer function:

$$\frac{Y(j\omega)}{E(j\omega)} = k_p \left\{ \frac{1+j\omega\tau_d}{1+j\omega\frac{\tau_d}{10}} \right\} \quad (14.48)$$

Make use of the approximation formula of (14.32).

15
Practical rules for direct digital control (DDC)

15.1 INTRODUCTION

In this chapter we shall discuss some options for choosing control parameters for direct digital control of a process. Together with parameters such as proportional gain, integration time, and differentiating time that also occur in continuous systems, we find two parameters that are specific for digital control, namely the sampling frequency and the resolution of the applied digital code. The latter parameter, the resolution, determines the accuracy with which the sampling values are converted into digital words. Also in calculation and in reconstruction, resolution plays an important part.

In the usual resolutions for AD and DA conversion of 8, 10, 12 or more bits the accuracy to be achieved is always so large that this property of digital control has hardly any negative effect on the control behaviour of the system.

The resolution is expressed in the magnitude and the degree of constancy of the controlled variable of the process. High resolution (many bits per digital word) has a favourable influence on this controlled value of the process.

As we have seen in the previous chapter, it is often necessary using more bits (double word length) for calculation purposes. In most front-ends this is adjustable.

15.2 CHOOSING A SAMPLING FREQUENCY

If we choose the sampling frequency:

$$f_s = \frac{1}{T} \text{ (Hz)} \tag{15.1}$$

it has a large influence on the dynamic behaviour of the controlled system. A larger value of the sampling period T results in larger phase lag in the sampling and reconstruction process, for this extra phase lag in a ZOH applies (see also section 10.7):

$$\varphi_h \simeq -\frac{1}{2}\omega T \qquad (15.2)$$

It is clear that increasing the phase lag is disadvantageous to the degree of stability of the system. A stable, controlled system may even become unstable by an increase of T.

The factors that determine the practical choice of T come from the digital part of the controller (the process computer) and the process to be controlled. If the process computer determines the choice of T, attention must be paid to the required calculation time for the algorithm and the conversion times of the ADC and the DAC.

As we have seen in section 14.5 it is also very important that we choose the not too high sampling frequency in connection with the accuracy of the algorithm calculations.

Example 15.1

A microcomputer is used for DDC of a number of identical control loops. The control algorithm requires the processing of 120 instructions; the conversion time of the ADC is 100 μs. Except those for controlling the loop, 40 instructions per loop are used for controlling devices, verifying the control loop, and data logging. Every instruction takes 2 μs. How many loops might be controlled by the microcomputer, if this number is totally defined by the possibilities of the control devices?

After analog digital conversion the calculation of the control algorithm may start. A new control magnitude will be calculated $(100 + 2 \times 120) = 340$ μs, after sampling, and will be fed to the input register of the DAC. The instructions for controlling, verifying, and data logging may fall within the conversion time of the ADC. After feeding the new control value to the DAC a new loop may be controlled. So $(1/340 \cdot 10^{-6}) \simeq 3000$ samplings per second may take place. Usually the properties of the process to be controlled will determine the choice of T. In general we may say that Shannon's sampling theorem must be satisfied. Because of the extra loss of information occurring in AD and DA conversion in the calculation procedures of the control algorithm, and especially in the hold device, we shall always choose a smaller value for T than that implied by Shannon's theorem, as we have seen in section 10.7.

A starting point for choosing the sampling frequency is offered by the highest signal frequency ω_h; this is the frequency (bandwidth) at which the modulus characteristic of the open loop system intersects the 0 dB axis. This frequency is also called the cross over frequency. A practical choice for the sampling frequency is provided by: $\omega_s = 5 \ldots 10 \; \omega_h$.

For a process in which one dominating time constant τ_1 of the controlled system occurs and in which a posssible time delay is many times smaller than τ_1, a practical choice for τ_1 is:

$$T \leqq \frac{\tau_1}{5} \qquad (15.3)$$

For processes in which new information is obtained by an inherent signal sampling

T_k, for example by quality analysis, the sampling frequency must be synchronized with the quality analysis. Then we have:

$$T = T_k \tag{15.4}$$

For such processes the result of the inherent sampling must be known before the process computer starts a new calculation.

Example 15.2
Let us assume that the microcomputer described in Example 15.1 is used for controlling a number of identical control loops. How many identical loops may be controlled if every loop has a dominant time constant of 0.5 s? According to (15.3) the choice for T is defined to be 100 milli seconds. In this time:

$$n = \frac{100 \times 10^{-3}}{340 \times 10^{-6}} \simeq 300 \tag{15.5}$$

control loops may be controlled.

How many identical loops may be controlled if every loop has quality analysis that lasts 5 s?

According to (15.4) we must choose $T = 5$s, with which the number of loops to be controlled would be:

$$n = \frac{5}{340 \times 10^{-6}} \simeq 14700$$

More often we shall find that there are one or at most a few loops with such a quality analysis, the others being faster loops.

15.3 ADJUSTMENT OF K_R, τ_i, AND τ_d

To adjust control parameters for continuous control systems we often use the adjustment rules of Ziegler and Nichols. The parallel controller which these rules are based on, has a transfer function:

$$H_R(s) = K_R \left(1 + \frac{1}{\tau_i s} + \tau_d s\right) \tag{15.6}$$

For a cascade controller, with the transfer function

$$H_R(s) = K_R' \left(1 + \frac{1}{\tau_i' s}\right)(1 + \tau_d' s) \tag{15.7}$$

the adjustment rules may also be used, if we take into account that the last controller may be written as in (15.6).

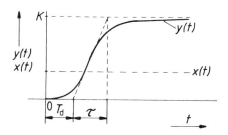

Fig. 15.1 — Approximation of a step response by those of a time delay and a first order system.

Processes showing a step response of the form in Fig. 15.1 may be described by the transfer function:

$$H_p = \frac{Y(s)}{X(s)} \simeq K \frac{e^{-T/d s}}{1+\tau s} \,. \qquad (15.8)$$

From this description it is assumed that the process is sufficiently characterized by a dc gain K, a time delay T_d and a first order system with time constant τ. For such processes we have, according to Ziegler and Nichols, the practical adjustment rules for K_R, τ_i, and τ_d, described in Table 15.1. It is assumed that the transfer function of the controller is (15.6).

If the effect of signal sampling and reconstruction of a sampled control loop is negotiated into an extra delay of $\tfrac{1}{2}T$, with regard to a continuous controlled process, than we may derive the rules for Table 15.1b from those according to Table 15.1a. The time delay of the process is now increased by $\tfrac{1}{2}T$. This approximation applies only if $T \ll T_d$.

Example 15.3
The step response of a (non-controlled) system is shown in Fig. 15.2.
Determine the control actions for a PID control algorithm.

Answer:
The process may be described by the transfer function:

$$H(s) = \frac{3 \, e^{-s}}{1 + 2s} \qquad (15.9)$$

A good choice for the sampling period is, for example, $T = 0.2$ s; this introduces an

Table 15.1 — Comparison of adjustment rules for continuous and sampled data systems

Controller	K_R	τ_i	τ_d
P	$\dfrac{\tau}{T_d} \cdot \dfrac{1}{K}$	—	—
PI	$0.9 \dfrac{\tau}{T_d} \cdot \dfrac{1}{K}$	$3.3\, T_d$	—
PID	$1.2 \dfrac{\tau}{T_d} \cdot \dfrac{1}{K}$	$2\, T_d$	$0.5\, T_d$

a Adjustment rules for continuous control systems

Controller	K_R	τ_i	τ_d
P	$\dfrac{\tau}{T_d + \frac{1}{2}T} \cdot \dfrac{1}{K}$	—	—
PI	$0.9 \dfrac{\tau}{T_d + \frac{1}{2}T} \cdot \dfrac{1}{K}$	$3.3\,(T_d + \tfrac{1}{2}T)$	—
PID	$1.2 \dfrac{\tau}{T_d + \frac{1}{2}T} \cdot \dfrac{1}{K}$	$2(T_d + \tfrac{1}{2}T)$	$0.5\,(T_d + \tfrac{1}{2}T)$

b Adjustment rules for sampled data systems.

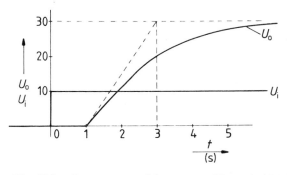

Fig. 15.2 — Step response of the process of Example 15.3.

extra time delay of 0.6 s. Then the recommended settings for the control actions may be read from Table 15.1b:

$$K_R = 1.2 \times \frac{2}{1.1} \times \frac{1}{3} = 0.7$$

$$\tau_i = 2(1+0.1) = 2.2 \text{ s} \tag{15.10}$$

$$\tau_d = 0.5(1+0.1) = 0.55 \text{ s}$$

By means of the methods discussed in Chapter 14 may we determine the transfer function $D(z)$ of the digital network, from (15.6) and (15.10).

A second possibility, presented by Ziegler and Nichols for setting controlled processes, is their procedure in which the controlled process is made to oscillate first. The controller again has the transfer function of (15.6). This procedure is also suitable for the digitally controlled process. Assuming a proportional controller, the proportionality factor is such that the system will start oscillating. Assume the then value of the proportional gain in the controller to be $K_{R,osc}$. The duration of the period of the then sinusoidally varying output signal is measured; assume this value to be T_{osc}. Table 15.2 indicates how, according to Ziegler and Nichols, a feasible

Table 15.2 — Adjustment rules by Ziegler and Nichols according to the oscillation method

Control actions	K_R	τ_i	τ_d
P	$0.5 K_{R,osc}$	—	—
PI	$0.45 K_{R,osc}$	$0.85 T_{osc}$	—
PID	$0.6 K_{R,osc}$	$0.5 T_{osc}$	$0.125 T_{osc}$

controlled system may be set. Note that K_R, τ_i, and τ_d must still be transformed into the digital control algorithm.

Example 15.4
To set the control parameters of a digitally PID controlled process, this process is made to oscillate. The gain factor appears to be 6, and the oscillation frequency

is 0.25 Hz. Starting from these data and the indicated values for K_R, τ_i, and τ_d in Table 15.2, we find the following settings:

$$K_R = 3.6; \tau_i = 2 \text{ s, and } \tau_d = 0.5 \text{ s.}$$

From these data we may determine the transfer function $D(z)$ of the digital control algorithm.

15.4 DIGITAL PROCESS CONTROL DESIGN IN THE BODE DIAGRAM

If we start from the extra time delay of $\frac{1}{2}T$ resulting from the sampling and reconstruction process, we may also use a design procedure in the Bode diagram to determine K_R, τ_i, and τ_d.

First, the Bode diagram of the process is modified in such a way that the extra delay $\frac{1}{2}T$ is integrated. Because:

$$\left| e^{-j\omega(T/2)} \right| = 1 \text{ and arg } e^{-j\omega(T/2)} = -\omega\frac{T}{2} \tag{15.11}$$

only the argument characteristic needs to be adapted. Thus we may apply a design procedure to the adapted Bode diagram.

Example 15.5

The Bode diagram of a process is shown in Fig. 15.3 We are asked to design a digital controller with a position PID control algorithm. For the present frequency band the process may be described by the transfer function:

$$H(j\omega) = \frac{10}{(1 + j\omega\frac{1}{4})(1 + j\omega\frac{1}{8})(1 + j\omega\frac{1}{20})} \tag{15.12}$$

From the Bode diagram it appears that $\omega_h \approx 15$ rad/s (please check!). A good choice for T is found from:

$$\omega = 10\omega_h \approx 150 \text{ rad/s}$$

so: $\quad T_s = 42 \text{ ms} \tag{15.13}$

The extra argument that arises from signal sampling and reconstruction becomes

$$\varphi = -\omega\frac{T_s}{2} = -\frac{\omega}{48} \text{ rad}$$

The broken line in the argument characteristic of Fig. 15.3 indicates the modified characteristic in which φ is integrated. By means of the adapted Bode diagram we

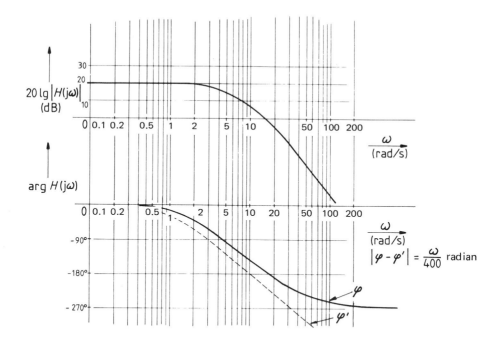

Fig 15.3 — Bode diagram for Example 15.5.

may subsequently define the control parameters K_R, τ_i and τ_d. For the procedure see ref. [8]. The control parameters obtained will have to be transformed into parameters for the digital control algorithm.

15.5 COMPARISON OF ANALOG AND DIGITAL PROCESS CONTROL

We shall demonstrate the differences between analog and digital control by means of some step responses. The process being considered has a transfer function:

$$H_1(s) = \frac{1}{(1+s)(1+\tfrac{1}{2}s)} \tag{15.14}$$

The requirements for the controlled process are:

1. The steady state error resulting from a step on the input must be zero.
2. The overshoot in the step response may be about 25%.

Let us first consider the solution with a continuous controller. The block diagram for this case is shown in Fig. 15.4.

To satisfy the first requirement we may choose a controller:

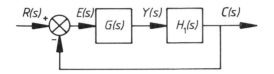

Fig. 15.4 — Block diagram of the continuous controlled process.

$$G(s) = K_R \left(1 + \frac{1}{s}\right) \text{ (PI-controller)} \tag{15.15}$$

The transfer function of the open loop system will be:

$$G(s) \cdot H(s) = K_R \left(1 + \frac{1}{s}\right) \cdot \frac{1}{(s+1)(\tfrac{1}{2}s+1)} = \frac{2K_R}{s(s+2)} \tag{15.16}$$

To satisfy the second requirement we shall consider the transfer function of the closed loop system:

$$\frac{C(s)}{R(s)} = \frac{GH}{1+GH} = \frac{2K_R}{s^2 + 2s + 2K_R} \tag{15.17}$$

If we compare (15.17) to the standard form, see Chapter 2, of a second order system we find:

$$\left. \begin{array}{l} \omega_0^2 = 2K_R \\ 2\beta\omega_0 = 2 \end{array} \right\} \tag{15.18}$$

For an overshoot in the step response of about 25%, the damping ratio β may be about:

$$\beta = 0.4 \tag{15.19}$$

With this information we find from (15.18):

$$K_R \simeq 3 \tag{15.20}$$

The transfer function of the analog controller now becomes:

Sec. 15.5] Comparison of analog and digital process control

$$G(s) = 3\left(1 + \frac{1}{s}\right) \quad (15.21)$$

Next we shall consider the solution with a digital controller. The block diagram will be shown in Fig. 15.5.

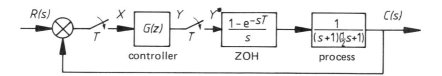

Fig. 15.5 — Block diagram of the digitally controlled process.

When realizing the digital controller we shall use the above values of the continuous controller setting. The transfer function of the digital PI controller becomes (position algorithm):

$$G(z) = K'_R\left(1 + \frac{K_i}{1 - z^{-1}}\right) \quad (15.22)$$

with $\quad K'_R = K_R$ and $K_1 = \dfrac{T}{\tau_i} = T$, for $\tau_i = 1$ s.

Choose from the sampling time $T = 0.1$ s. For $G(z)$ this yields:

$$G(z) = 3\left(1 + \frac{0.1}{1 - z^{-1}}\right) = 3\left(\frac{1.1 - z^{-1}}{1 - z^{-1}}\right) = \frac{Y(z)}{X(z)} \quad (15.23)$$

In the time domain this yields:

$$y(nT) = y(nT - T) + 3.3x(nT) - 3x(nT - T) \quad (15.24)$$

The direct programming of this controller is shown in Fig. 15.6.

Both control systems are simulated; the step responses of the system with continuous and digital controller are shown in Fig. 15.7. The overshoot in both responses is, as was to be expected, about 25%. The difference between the responses is nil; the digital control yields a little more overshoot, resulting from the sampling and reconstruction process.

In Fig. 15.8 we have step responses for various sampling periods: $T = 0.05$,

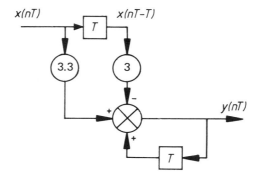

Fig. 15.6 — Realization diagram of the PI controller with position algorithm.

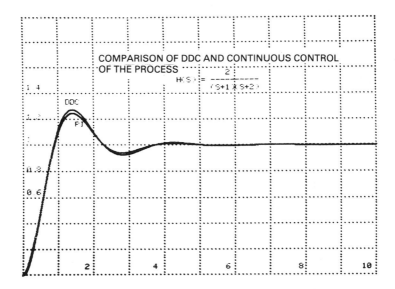

Fig. 15.7 — Step responses of analog and digitally controlled process.

$T = 0.1$, and $T = 0.5$ s. Of course the controller settings have been adapted. It is clear that the control worsens, as the sampling time increases, as has been stated with the derived algorithms.

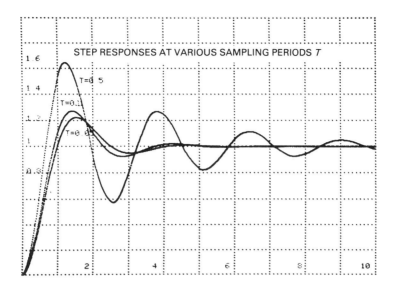

Fig. 15.8 — Step responses for different values of T.

15.6 PROBLEMS

1. Given the continuous process control according to Fig. 15.9.

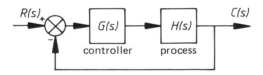

Fig. 15.9 — Continuous process control according to Problem 1.

The transfer function of the process is:

$$H(s) = \frac{1}{(s+1)(2s+1)(3s+1)} \tag{15.25}$$

$G(s)$ is the transfer function of a parallel PID controller, designed according to the rules of Ziegler and Nichols (oscillation method).

a Determine the transfer function $G(z)$. (Note: First determine K_{osc} and T_{osc}!).
b Determine, from the result obtained in (a), the difference equation of the

digital realization of the PID controller (position algorithm). Choose for the sampling period $T = 0.1$ s.
2. Given the unit step response, see Fig. 15.10, of the process according to (15.25).

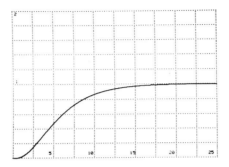

Fig. 15.10 — Step response of the system according to (15.25).

Choose T and determine the setting of the PID controller, but now by the step response method of Ziegler and Nichols. Also determine the realization diagram if the controller is a digital one.
3. A first order process with a dc gain of 10 and time constant of 10 seconds is digitally controlled by a PI controller with a position algorithm. Choose a value for T and determine $D(z)$ of the controller, if the bandwidth of the controlled system must be 1 rad/s. Repeat this for a PI controller with a velocity control algorithm.

Note: Determine ω_h and then choose an appropriate value of ω_s. Then choose the I action in such a way that a feedback integrator develops.

16

Time-optimal control systems

16.1 INTRODUCTION

In modern control engineering optimal control criteria are used more and more in designing systems. By doing so we try to minimize a specific (error) performance index. One or more quantities, such as fuel consumption and the process time, must be kept as small as possible. If we consider time to be a performance index we speak of a time-optimal control system.

In the theories on continuous linear systems such a control system comes down to an on/off control, in which a (maximum) positive and/or negative control signal is supplied to the system for certain periods (this is also called 'bang-bang control').

The difference between this system and sampled data systems is clear, the control signals for the process can be supplied only at the sampling instants, so the process may be controlled only at fixed times. Special design methods have been developed, of which we shall discuss one: the 'minimum settling time control', also called 'dead-beat control'.

16.2 THE DEAD-BEAT RESPONSE METHOD

Our starting point for dead-beat control is the block diagram of Fig. 16.1; the digital

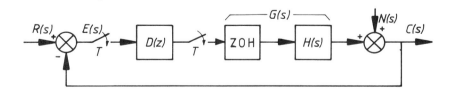

Fig. 16.1 — Block diagram of process control in which $D(z)$ is designed according to the dead-beat response method.

controller $D(z)$ must be designed in such a way that the system may be transferred from one situation to another as soon as possible.

We must distinguish between servosystems and regulator systems. In servosystems the output $c(nT)$ must follow an input change Δr as soon and as well as possible. The requirements as to the control system are:

a. The steady state error $e(\infty)$ must be equal to zero and must also be zero as soon as possible at the sampling instants; that is, $e(nT) = 0$ for minimum value of n.
b. It must be possible to realize $D(z)$.

The latter seems trivial, but we shall see that this requirement determines the design of $D(z)$ to a large extent.

The first requirement — $e(nT) = 0$ as soon as possible — may be formulated differently:

$$c(nT) = r(nt) \text{ for minimum value of } n \qquad (16.1)$$

The transfer function of the controlled system is assumed to be $M(z)$, so that:

$$\frac{C(z)}{R(z)} = M(z) = \frac{D(z)G(z)}{1 + D(z)G(z)} \qquad (16.2)$$

After transformation there follows for the controller $D(z)$:

$$D(z) = \frac{M(z)}{1 - M(z)} \cdot \frac{1}{G(z)} \qquad (16.3)$$

and for the error signal $E(z)$:

$$E(z) = R(z)[1 - M(z)] \qquad (16.4)$$

In general, $M(z)$ will be finite polynomial in z, according to:

$$M(z) = m_0 + m_1 z^{-1} + m_2 z^{-2} + \ldots \qquad (16.5)$$

As every physical system has more poles than zeros in the transfer function, an input change will not be noticeable at the output directly. In a sampled data system this will happen after at least one sampling period. From this it follows that m_0 in expression (16.5) must be equal to zero for $M(z)$, so that:

$$M(z) = m_1 z^{-1} + m_2 z^{-2} + \ldots \qquad (16.6)$$

The requirement that the output follows the input as soon as possible (at the sampling instants) is satisfied by taking into formula (16.6) as few terms m_i ($i = 1, 2, \ldots$) as possible. Likewise, if the process to be controlled contains a time delay equal to k times the sampling period, an output change will be measurable only

The dead-beat response method

after kT seconds, so that at least one term z^{-k} must be taken into account for $M(z)$ in expression (16.6). In any case, according to the first requirement (a), we must have:

$$\lim_{n\to\infty} e(nT) = \lim_{z\to\infty} \frac{z-1}{z} E(z) = \lim_{z\to 1} \frac{z-1}{z} R(z)[1-M(z)] = 0 \qquad (16.7)$$

From this it follows that $M(z)$ is also determined by $R(z)$. According to formula (16.3) this results in $D(z)$ being determined by $R(z)$ too. This consequence, namely that the design of the controller, $D(z)$, depends on the input signal, is one of the disadvantages of dead-beat control.

A special case occurs if the input changes as a step; then for $e(\infty)$ formula (16.4) and $R(z) = \frac{z}{z-1}$):

$$e(\infty) = \lim_{z\to 1} \frac{z-1}{z} E(z) = \lim_{z\to 1} \frac{z-1}{z} \cdot \frac{z}{z-1} [1-M(z)] = 0 \qquad (16.8)$$

of:

$$\lim_{z\to 1} [1-M)z)] = 0 \qquad (16.9)$$

This is achieved by choosing, for $M(z)$:

$$M(z) = \frac{1}{z^n} \qquad (16.10)$$

in which we have $m_n = 1$ and $m_i = 0$ for $i \neq n$.

So the output is also (at the sampling instants) a step signal, but shifted by nT seconds. Then the controlled system has an n-fold pole in the origin of the z-plane.

Substitution of (16.10) into (16.3) then yields, for the controller $D(z)$:

$$D(z) = \frac{z^{-n}}{1-z^{-n}} \cdot \frac{1}{G(z)} = \frac{1}{z^n - 1} \cdot \frac{1}{G(z)} \qquad (16.11)$$

From the above and from the examples in the next section it appears that the minimal value of n in expressions (16.10) and (16.11) is detemined by the time delay in the process to be controlled. It is considerably more difficult to determine $M(z)$ for other input signals. Assume, for example, the input to be a ramp; now we have:

$$R(z) = \frac{zT}{(z-1)^2} \qquad (16.12)$$

Then for $e(\infty)$:

$$e(\infty) = \lim_{z \to 1} \frac{z-1}{z} R(z) [1 - M(z)] =$$

$$= \lim_{z \to 1} \frac{z-1}{z} \cdot \frac{zT}{(z-1)^2} [1 - M(z)] =$$

$$= T \lim_{z \to 1} \frac{1}{z-1} [1 - M(z)] = 0 \qquad (16.13)$$

To obtain this the term $1 - M(z)$ must contain at least one factor $(z-1)^2$. Of $M(z)$ at least two terms must be taken into account. For convenience we assume the time delay in the process to be equivalent to zero and assume, for $M(z)$:

$$M(z) = m_1 z^{-1} + m_2 z^{-2} \qquad (16.14)$$

Thus we obtain:

$$1 - M(z) = f(z)(z-1)^2 \qquad (16.15)$$

or:

$$1 - \frac{m_1 z + m_2}{z^2} = f(z)(z^2 - 2z + 1) \qquad (16.16)$$

with:

$$\lim_{z \to 1} f(z) = \text{constant} \qquad (16.17)$$

Now follows:

$$\frac{z^2 - m_1 z - m_2}{z^2} = f(z)(z^2 - 2z + 1)$$

Choose $m_1 = 2$ en $m_2 = -1$ so that

$$M(z) = 2z^{-1} - z^{-2} \qquad (16.18)$$

This represents a more complex relation between the input and output of the system than in case of a step response. Then $D(z)$ may be determined from formulas (16.18) and (16.3).

In a regulator system the output $c(nT)$ must be kept as constant as possible, under the influence of a disturbance N which is put onto the output in the block diagram of Fig. 16.1. Here:

a $c(nT) = 0$, for minimal value of n. Here we assume $r = 0$
b It must be possible to realise $D(z)$.

If we assume the transfer function of the disturbance behaviour $\dfrac{C(z)}{N(z)} = M'(z)$ and the controller $D'(z)$ to be designed, $(R(z) = 0)$:

$$\frac{C(z)}{N(z)} = M'(z) = \frac{1}{1 + D'(z)G(z)} \tag{16.19}$$

or:

$$D'(z) = \frac{1 - M'(z)}{M'(z)} \cdot \frac{1}{G(z)} \tag{16.20}$$

Similar considerations as those for the servosystems apply for $M'(z)$ and $D'(z)$. Comparison to formula (16.3) shows that $M'(z) = 1 - M(z)$.

16.3 APPLICATIONS

By means of two examples the dead-beat response method will be illustrated more closely.

Example 16.1
Given the block diagram according to Fig. 16.2; $T = 1$ s.

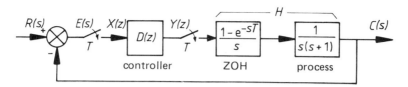

Fig. 16.2 — Block diagram for Example 16.1.

Input R changes by steps. We are to design $D(z)$ according to the dead-beat response method. The following requirements must be satisfied:

a $\lim\limits_{t \to \infty} e(t) = 0$ or $\lim\limits_{z \to 1} \dfrac{z-1}{z} E(z) = 0$.

b $c(nT) = r(nT)$ or $C(z) = z^{-n} R(z)$, for minimal value of n.

c It must be possible to realise $D(z)$.

Solution:

$$\frac{C(z)}{R(z)} = \frac{D(z)H(z)}{1+D(z)H(z)} = z^{-n} \tag{16.21}$$

so

$$D(z)H(z) = z^{-n} + z^{-n}D(z)H(z)$$

From this we obtain:

$$D(z) = \frac{z^{-n}}{1-z^{-n}} \cdot \frac{1}{H(z)}$$

or:

$$D(z) = \frac{1}{(z^n - 1)H(z)} \tag{16.22}$$

For $H(z)$:

$$H(z) = \mathscr{Z}\left[\frac{1-e^{-sT}}{s} \cdot \frac{1}{s(s+1)}\right]$$

$$= (1-z^{-1})\,\mathscr{L}\left[\frac{1}{s^2(s+1)}\right]$$

$$= (1-z^{-1})\,\mathscr{L}\left[\frac{1}{s^2} - \frac{1}{s} + \frac{1}{s+1}\right]$$

$$= \frac{z-1}{z}\left[\frac{zT}{(z-1)^2} - \frac{z}{z-1} + \frac{z}{z-e^{-T}}\right]$$

$$= \frac{z(T+e^{-T}-1) - Te^{-T} - e^{-T} + 1}{(z-1)(z-e^{-T})} \tag{16.23}$$

With $T=1$ and $e^{-1} \simeq 0.37$:

$$H(z) = \frac{0.37z + 0.26}{(z-1)(z-0.37)} \tag{16.24}$$

Substitution into formula (16.22) yields:

$$D(z) = \frac{(z-1)(z-0.37)}{(z^n - 1)(0.37z + 0.26)}$$

Because it must be possible to realise $D(z)$ (order numerator $\leqq$ denominator), we may choose $n = 1$. For $D(z)$:

$$D(z) = \frac{(z-1)(z-0.37)}{(z-1)(0.37z + 0.26)} = \frac{z - 0.37}{0.37z + 0.26}$$

or:

$$D(z) = \frac{1 - 0.37z^{-1}}{0.37 + 0.26z^{-1}} \qquad (16.25)$$

Realization of $D(z) = \dfrac{Y(z)}{X(z)}$ yields:

$$0.37Y(z) + 0.26z^{-1}Y(z) = X(z) - 0.37z^{-1}X(z) \qquad (16.26)$$

In the t-domain this yields with aid of the shift rule:

$$0.37y(nT) + 0.26y(nT - T) = x(nT) - 0.37x(nT - T)$$

or:

$$y(nT) = 2.72x(nT) - x(nT - T) - 0.71y(nT - T) \qquad (16.27)$$

The realization diagram (direct programming) of formula (16.27) is represented in Fig. 16.3.

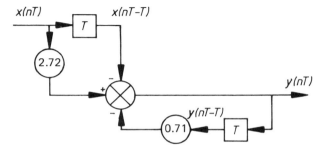

Fig. 16.3 — Realization diagram of the dead-beat controller of Example 16.1.

By choosing $n = 1$ and the resulting transfer function $D(z)$ we obtain:

$$\frac{C(z)}{R(z)} = z^{-1} \tag{16.28}$$

The output $c(nT)$ will follow $r(nT)$ if this is a step, after one sampling period (at least at the sampling instants!).
For $E(z)$:

$$E(z) = R(z) - C(z) = R(z) - z^{-1} R(z) = (1 - z^{-1}) R(z) \tag{16.29}$$

Since $r(t)$ is a unit step, we find $R(z) = \dfrac{z}{z-1}$ so that:

$$E(z) = \frac{z-1}{z} \cdot \frac{z}{z-1} = 1 \tag{16.30}$$

So during the first sampling period the error signal $e(nT)$ will have a value of 1 and thereafter it will be equal to zero.

The responses $c(t)$ and $e(t)$ are represented in Figs 16.4 and 16.5 respectively.

Note that after one sampling period (1 s), the output is indeed equivalent to 1 at the sampling instants. Because of round-off errors ($e^{-1} = 0.367879 \ldots$) this is not exact.

Also pay attention to the so called intersample ripple between the sampling instants!

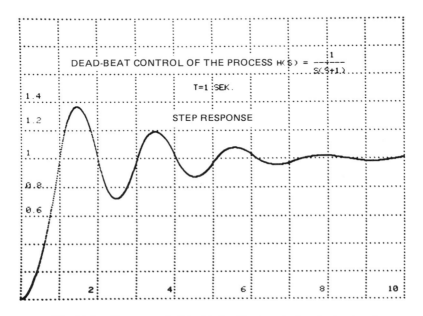

Fig. 16.4 — Response of $c(T)$ with dead-beat control and a step input.

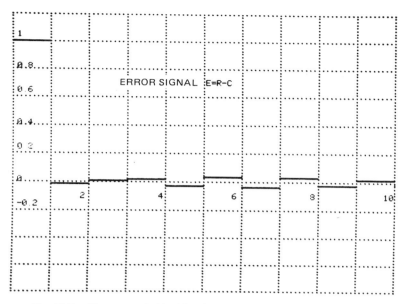

Fig. 16.5 — Response of e(t) with a dead-beat control and a step input.

Example 16.2
Given the controlled process according to Fig. 16.6; again T is 1 s.

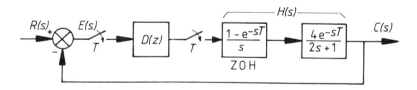

Fig. 16.6 — Block diagram for Example 16.2.

Again we consider a step response! $D(z)$ becomes:

$$D(z) = \frac{1}{(z^n - 1)H(z)} \text{ for minimum value of } n \qquad (16.31)$$

Here:

$$H(z) = \mathscr{L}\left[\frac{1 - e^{-sT}}{s} \cdot \frac{4e^{-sT}}{2s+1}\right]$$

$$= 4(1-z^{-1})z^{-1} \mathscr{L}\left[\frac{1}{s(2s+1)}\right]$$

$$= \frac{4(z-1)}{z^2} \mathscr{L}\left[\frac{1}{s} - \frac{1}{s+\frac{1}{2}}\right]$$

$$= \frac{4(z-1)}{z^2} \cdot \left[\frac{z}{z-1} - \frac{z}{z-e^{-T/2}}\right]$$

$$= \frac{4(1-e^{-\frac{1}{2}T})}{z(z-e^{-\frac{1}{2}T})} \tag{16.32}$$

With $T=1$ and $e^{-\frac{1}{2}} = 0.607$ follows:

$$H(z) = \frac{1.57}{z(z-0.607)} \tag{16.33}$$

From this:

$$D(z) = \frac{1}{z^n - 1} \cdot \frac{z(z-0.607)}{1.57} \tag{16.34}$$

Because $D(z)$ must be realizable, we have in this case the lowest value for n, $n=2$. Numerator and denominator will both be of the second degree.

After division by z^2 this yields for $D(z)$:

$$D(z) = \frac{1 - 0.607 z^{-1}}{1.57(1 - z^{-2})} = \frac{Y(z)}{X(z)} \tag{16.35}$$

In the time domain this yields:

$$1.57 y(nT) - 1.57 y(nT - 2T) = x(nT) - 0.607 x(nT - T)$$

or:

$$y(nT) = 0.635 x(nT) - 0.385 x(nT - T) + y(nT - 2T) \tag{16.36}$$

The realization diagram (direct programming) is shown in Fig. 16.7.

By choosing $D(z)$ and $n=2$ the transfer function of the controlled system becomes:

$$\frac{C(z)}{R(z)} = z^{-2} \tag{16.37}$$

Sec. 16.4]   **Appreciation of dead-beat control systems**   305

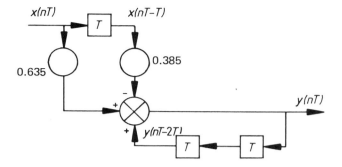

Fig. 16.7 — Realization diagram of the dead-beat controller of Example 16.2.

Fig. 16.8 shows the step response $c(t)$. Only after two sampling periods (2 s) output $c(nT)$ will follow the input at the sampling instants. In this case too, the following will not be exact, because of round-off errors. In this case we have no intersample ripple.

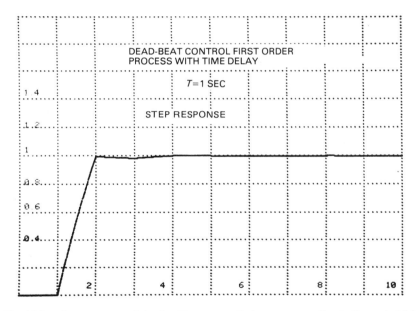

Fig. 16.8 — Step response of the dead-beat controlled system according to Example 16.2.

16.4 APPRECIATION OF DEAD-BEAT CONTROL SYSTEMS

Although, at first, dead-beat control systems seem to function well if we consider time to be a performance index, the method has a number of drawbacks, because of which this method is less popular nowadays.

From the previous section it appeared that the controller to be designed, $D(z)$, not only depends on the properties of the process, but that it is also a function of the character of the input and/or the error signal.

The design of $D(z)$ also depends on the sampling time T, and is sensitive to parameter variations in the process. For from (16.3) it appears that the reciprocal transfer function $(1/G(z))$ becomes part of $D(z)$. Thus a variation in a pole and/or zero of $G(z)$ will influence the effect of a dead-beat control adversely (no more exact! following at the sampling instants).

Generally, because $D(z)$ has the zeros of $G(z)$ as its poles, oscillation will occur in the control signal after feedback. The hold device makes sure that it has a pulse-like character. As these control singles change at the sampling instants, the frequency in the output signal will be half the sampling frequency. This is shown in Fig. 16.9.

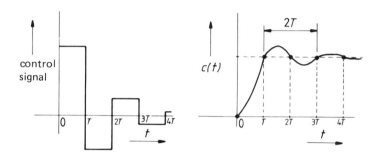

Fig. 16.9 — Intersample ripple occurring in the output signal.

The oscillation occurring in the output signal between the sampling instants is called intersample ripple; it is also one of the disadvantages of dead-beat control. Note that the output signal shows no oscillation at the sampling instants $c(nT)$. A possible way to avoid intersample ripple is the so called ripple free response, which is determined by means of the modified z-transformation and therefore lies outside the scope of this book.

16.5 PROBLEMS

1 Given: the block diagram of a sampled data system according to Fig. 6.10. We want to design the digital network $D(z)$ in such a way that the output at the sampling instants follows a step change at the input as soon as possible.

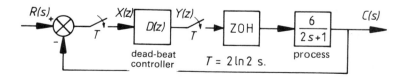

Fig. 16.10 — Block diagram of the dead-beat control system of Problem 1.

a Determine the transfer function $D(z)$.
b Draw the realization diagram of $D(z)$.
2 Given: a system with z-transformed transfer function:

$$H(z) = \frac{K(3z-1)}{3(z-1)(z-\tfrac{1}{2})}$$

The system has unity feedback.
a Explain by means of a root locus construction that dead-beat response develops for a certain value of K.
b Determine the value for K as meant in (a).
c Sketch the step response. Explain the overshoot in the response.

17
Digital filters

17.1 INTRODUCTION

In practice continuous signals are always affected by noise. From theories on continuous systems we know that this noise is usually filtered (weakened) sufficiently by the low-frequent character of that system (controller plus process).

This is different for sampled data systems. On the one hand, higher harmonics are introduced into the system by the sampling process, whereas on the other hand high-frequency components get into the system because of the noise already present. In the first case the higher harmonics will be filtered out, as in continuous systems, by the low frequency character of the hold device and the process. In the second case the high-frequency noise components will get into the sampled signal as low-frequency components. Fig. 17.1 shows the worst case; here the noise band is near the

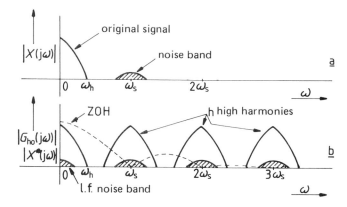

Fig. 17.1 — The effect of a noise band in the continuous signal on the frequency spectrum of the sampled signal (worst case).

sampling frequency ω_s. In this figure we also find the filtering action (frequency spectrum) of a zero order hold device.

Fig. 17.1 shows that the high-frequency noise components are completely within the low-frequency spectrum of the original signal. Filtering by ZOH will no longer be possible.

If the noise band is not near ω_s, the situation will improve a little because the noise band is outside the low-frequency spectrum of the original signal as a result of sampling; then there is some filtering by the hold device.

From the time-domain point of view it means that peaks in the signal (because of noise) occur more or less at the sampling instants, which leads to errors in the sampled signal.

It is clear that this noise contribution must be reduced as much as possible in one way or another. In principle, two methods are applicable:

a Increasing the sampling frequency
b Analog filtering before sampling or digital filtering directly after sampling (anti-aliasing filter)

17.2 FILTERING BY INCREASING THE SAMPLING FREQUENCY

By increasing the sampling frequency with regard to the situation shown in Fig. 17.1 a better filtering of the noise band will develop. This is shown in Fig. 17.2. Because of

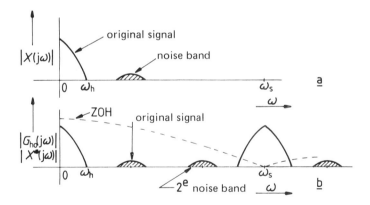

Fig. 17.2 — Effects of a higher sampling frequency on the noise filtering.

the sampling a second noise band will develop, with noise frequencies lower than the sampling frequency. Here we see that the noise bands are attenuated by the filtering of the zero order hold device, and that no low-frequency noise band will develop within the original signal band.

In any case, the increase of the sampling frequency must make sure that the second noise band (see Fig. 17.2) does not get into the original signal spectrum.

From a practical point of view we have seen that increasing the sampling frequency is not always possible or even undesirable.

17.3 SIGNAL FILTERING BEFORE SAMPLING

If the sampling frequency may not be raised sufficiently, it will be possible to filter the signal before sampling takes place. Such a filter will of course, be realized in the analog form.

The simplest analog filter is an RC network as in Fig. 17.3; its transfer function is:

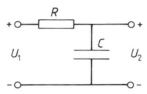

Fig. 17.3 — A simple *RC* filter.

$$G(j\omega) = \frac{U_2(j\omega)}{U_1(j\omega)} = \frac{1}{1 + j\omega\tau_f} \quad \text{with } \tau_f = RCs \qquad (17.1)$$

To investigate its filtering, we consider a noise component with frequency ω_s (worst case). The multiplication factor for this component because of the filter is:

$$|G(j\omega_s)| = \frac{1}{\sqrt{\omega_s^2 \cdot \tau_f^2 + 1}} \qquad (17.2)$$

If we require the attenuation of this component by means of the filter to be at least a factor 10, we have:

$$\frac{1}{\sqrt{\omega_s^2 \cdot \tau_f^2 + 1}} \simeq \frac{1}{10}$$

From this follows:

$$\omega_s \tau_f \simeq 10$$

Or:

$$\tau_f \simeq 10/\omega_s = \omega T/2\pi \simeq 1.6T \qquad (17.3)$$

So the time constant of the *RC*-filter must be at least 1.6 times the sampling period. However, the signal itself is also attenuated by the filter. Assume the sampling frequency to be 10 times as large as the highest occurring signal frequency ω_h, then the multiplication factor of the filter for this frequency is:

$$|G(j\omega_h)| = \frac{1}{\sqrt{\omega_h^2 \cdot \tau_f^2 + 1}} = \frac{1}{\sqrt{\frac{\omega_s^2}{100} \cdot \frac{100}{\omega_s^2} + 1}} = \frac{1}{\sqrt{2}} \triangleq -3\,\text{dB} \qquad (17.4)$$

Thus signal frequencies of ω_h are attenuated by 3 dB; of course, this will be less for lower signal frequencies.

Of course more complex filters may be applied (higher order filters). However, we must remember that increasing the filter action by higher order filters also introduces extra phase lag.

17.4 SIGNAL FILTERING AFTER SAMPLING

If a digital filter is applied, filtering takes place immediately after signal sampling. Of course the algorithm of such a filter may be combined with that of the control algorithm. No extra hardware will be needed; this situation is shown in Fig. 17.4.

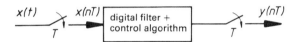

Fig. 17.4 — Digital filter combined with control algorithm.

The two most widely known digital filters are:

a exponential smoothing filter
b moving average filter

(a)
The algorithm for an exponential smoothing filter with input $x(nT)$ and ouput $y(nT)$ is:

$$y(nT) = \alpha x(nT) + (1-\alpha)y(nT - T) \qquad (17.5)$$

with $0 < \alpha < 1$. From this follows the difference equation:

$$y(nT - T) = \alpha x(nT - T) + (1-\alpha)y(nT - 2T) \qquad (17.6)$$

Substitution of (17.6) into (17.5) yields:

$$y(nT) = \alpha x(nT) + (1-\alpha)[\alpha x(nT-T) + (1-\alpha)y(nT-2T)] \qquad (17.7)$$

In the same way, $y(nT-2T)$ may be determined from (17.5) and substituted into (17.7). Then for $y(nT-2T)$ follows:

$$y(nT-2T) = \alpha x(nT-2T) + (1-\alpha)y(nT-3T) \qquad (17.8)$$

Substitution of (17.8) into (17.7) yields:

$$y(nT) = \alpha x(nT) + (1-\alpha)[\alpha x(nT-T) + (1-\alpha)[\alpha x(nT-2T) + \\ + (1-\alpha)y(nT-3T)]] \qquad (17.9)$$

In this equation we may again write $y(nT-3T)$ as a function of $x(nT-3T)$, etc. So in general for $y(nT)$ an infinite series will develop according to:

$$y(nT) = \alpha x(nT) + \alpha(1-\alpha)x(nT-T) + \alpha(1-a)^2 x(nT-2T) + \\ + \alpha(1-\alpha)^3 x(nT-3T) + \ldots \qquad (17.10)$$

If, in this last relation, we choose α to be, for example 0.5, then:

$$y(nT) = 0.5x(nT) + 0.25x(nT-T) + 0.125x(nT-2T) + \\ + 0.0625x(nT-3T) + \ldots$$

The contribution of $x(nT-kT)$ decreases exponentially for increasing values of k, hence the name exponential smoothing filter. We may also say that the sampling values are weighted with a weighting factor that decreases exponentially as the information gets 'older'. The general relation for $y(nT)$ according to (17.10) may be written as:

$$y(nT) = \alpha \sum_{k=0}^{\infty} (1-\alpha)^k x(nT-kT) \qquad (17.11)$$

After z-transform:

$$Y(z) = \alpha \sum_{k=0}^{\infty} (1-\alpha)^k z^{-k} X(z) \qquad (17.12)$$

which represents a geometrical series. Then for the transfer function of filter:

$$\frac{Y(z)}{X(z)} = \frac{\alpha}{1-(1-\alpha)z^{-1}} = \frac{z\alpha}{z-(1-\alpha)} \qquad (17.13)$$

Note:
The transfer function of (17.13) may also be determined directly from (17.5).

The transfer function of the analog RC network is:

$$\frac{Y(s)}{X(s)} = \frac{1}{1+s\tau_f} = \frac{1}{\tau_s}\frac{1}{s+\frac{1}{\tau_f}} \qquad (17.14)$$

After z-transform:

$$\frac{Y(z)}{X(z)} = \frac{1}{\tau_f}\frac{z}{z-e^{-T/\tau_f}} \qquad (17.15)$$

Comparison of expressions (17.13) and (17.15) yields, except for a constant in the transfer function:

$$1-\alpha = e^{-T/\tau_f}$$

or:

$$\tau_f = \frac{T}{\ln\left(\frac{1}{1-\alpha}\right)} \qquad (17.16)$$

For signals with little noise we choose α about one; for signals with much noise we choose α to be small.
Note that the original signal itself is attenuated more as α is assumed to be smaller.

(b)
A moving average filter gives an average of the last N signal values. Here:

$$\left.\begin{array}{ll} N = 1: & y(nT) = x(nT) \\ N = 2: & y(nT) = \tfrac{1}{2}[x(nT)+x(nT-T)] \\ N = 3: & y(nT) = \tfrac{1}{3}[x(nT)+x(nT-T)+x(nT-2T)] \end{array}\right\} \qquad (17.17)$$

etc.

After a transient period of $N-1$ values, an average of the last N sampling values is taken.

In general:

$$y(nT) = \frac{1}{N}[x(nT) + x(nT-T) + x(nT-2T) + \ldots + x(nT-(N-1)T)] \tag{17.18}$$

After z-transform:

$$Y(z) = \frac{1}{N}[X(z) + z^{-1}X(z) + \ldots + z^{-(N-1)}X(z)] \tag{17.19}$$

Or:

$$\frac{Y(z)}{X(z)} = \frac{1}{N}[1 + z^{-1} + \ldots + z^{-N} \cdot z^{+1}]$$

$$= \frac{1}{N}z[z^{-1} + z^{-2} + \ldots + z^{-N}] \tag{17.20}$$

The part between brackets represents a finite geometrical series, so that we have:

$$\frac{Y(z)}{X(z)} = \frac{1}{N}z \, z^{-1}\frac{1-z^{-N}}{1-z^{-1}} = \frac{1}{N} \cdot \frac{1-z^{-N}}{1-z^{-1}} = \frac{1}{N}\frac{z(z^N-1)}{z^N(z-1)} \tag{17.21}$$

Formula (17.21) represents the general transfer function of an N-order moving average filter.

It may be said that a moving average filter belongs to the class of transversal filters, which all have the property of the output being only a function of input values.

An additional property of a moving average filter is the fact that it is possible to filter out certain signal frequency components completely. From formula (17.21) follows the transfer function of the filter:

for $N = 2$: $\quad \dfrac{Y(z)}{X(z)} = \dfrac{1}{2} \cdot \dfrac{z(z^2-1)}{z^2(z-1)} = \dfrac{1}{2}\dfrac{(z-1)(z-1)}{z(z-1)} = \dfrac{1}{2}\dfrac{z+1}{z}$ (17.22)

for $N = 3$: $\quad \dfrac{Y(z)}{X(z)} = \dfrac{1}{3} \cdot \dfrac{z(zx^3-1)}{z^3(z-1)} = \dfrac{1}{3} \cdot \dfrac{(z^2+z+1)(z-1)}{z^2(z-1)} =$

Sec. 17.4] Signal filtering after sampling

$$= \frac{1}{3} \cdot \frac{z^2+z+1}{z^2} \tag{17.23}$$

for $N = 4$:
$$\frac{Y(z)}{X(z)} = \frac{1}{4}\frac{z(z^4-1)}{z^4(z-1)} = \frac{1}{4}\frac{z(z^2-1)(z^2+1)}{z^4(z-1)} =$$

$$= \frac{1}{4}\frac{(z+1)(z^2+1)}{z^3} \tag{17.24}$$

According to the z-transformation table we have, for a sinusoidal signal, the transfer function:

$$F(z) = \frac{T(z)}{z^2 - 2z\cos\omega T + 1} \tag{17.25}$$

in which $T(z)$ depends on the phase in $f(t)$. If we consider the signal frequencies $\frac{\omega_s}{2}$, $\frac{\omega_s}{3}$, and $\frac{\omega_s}{4}$ successively, (17.25) changes into:

$$F(z) = \frac{T(z)}{z^2 + 2z + 1} \tag{17.26}$$

in which for $\omega = \frac{\omega_s}{2}$, $T(z)$ can be divided by $z + 1$. From this, (17.26) changes into:

$$F(z) = \frac{T'(z)}{z+1} \quad \left(\omega = \frac{\omega_s}{2}\right) \tag{17.27}$$

for $\omega = \frac{\omega_s}{3}$:

$$F(z) = \frac{T(z)}{z^2 + z + 1} \quad \left(\omega = \frac{\omega_s}{3}\right) \tag{17.28}$$

for $\omega = \frac{\omega_s}{4}$:

$$F(z) = \frac{T(z)}{z^2 + 1} \quad \left(\omega = \frac{\omega_s}{4}\right) \tag{17.29}$$

Comparing formulas (17.22) to (17.24) to the formulas (17.27) to (17.29) shows that the poles of $F(z)$ are compensated by the zeros in the transfer function $\frac{Y(z)}{X(z)}$.

Conclusion
A moving average filter with $N = 2$ filters signal frequencies $\omega_s/2$ completely; a filter with $N = 3$ the signal frequencies $\omega_s/3$; a filter with $N = 4$, the signal frequencies $\omega_s/4$, etc.

In section 18.5 an application is shown.

17.5 PROBLEMS

1 Given; a digital filter which has a transfer function $D(z)$, input $E(z)$ and output $Y(z)$. The filter is part of sampled data system with sampling period $T = \ln 2$ s. The relation between $y(nT)$ and $e(nT)$ is given by the following difference equation:

$$y(nT) = \tfrac{3}{4} e(nT) + \tfrac{1}{4} y(nT - T)$$

The same filtering may be achieved by means of the RC network according to Fig. 17.5.

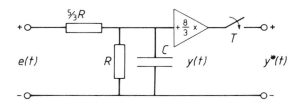

Fig. 17.5 — RC network of Problem 1.

Determine the value of the capacitor C of the analog filter, if $R = 1\,\text{M}\Omega$.
2 The realization diagram of a digital filter is shown in Fig. 17.6.

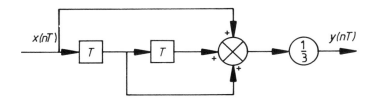

Fig. 17.6 — Realization diagram of Problem 2.

Determine the step response of the filter.

3 Given: the block diagram of Fig. 17.7.

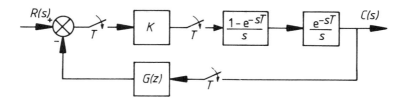

Fig. 17.7 — Block diagram of Problem 3.

The sampling period T is $T = 1$ s.

a Determine the positive values of K, for which the system is unstable, if $G(z)$ represents the transfer function of a second order moving average filter.
b Repeat (a), but now if $G(z)$ represents the transfer function of an exponential smoothing filter with $\alpha = 0.5$.

18

Case studies of digital control systems

18.1 INTRODUCTION

In this chapter we shall discuss three examples of sampled data systems. The first is a phase lock loop, in which we use a sample and hold device as a phase detector. In the second we shall discuss a temperature control, in which a digital PI controller is realized. The third example deals with a moving average filter.

In the last section the hardware of a digital controller will be explained, and the flow diagram for the software will be discussed.

18.2 PHASE LOCK LOOP WITH SAMPLE AND HOLD DEVICE

In the applications of Chapter 8 we have already discussed the phase lock loop as a continuous control system. The phase detector in the system was assumed to be continuous and linear. Usually, however, we use a sample and hold device (SH) as a phase detector. This is illustrated in Fig. 18.1.

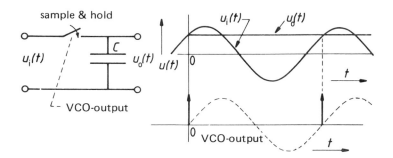

Fig. 18.1 — Use of a sample and hold device as a phase detector.

The input signal is sampled and held during the sampling period by a zero order hold device. The sampling frequency is equal to the frequency of the output signal of the voltage controlled oscillator (VCO). The instant (and so the phase) of sampling

Phase lock loop with sample and hold device

determines the output of the SH. Sampling may take place at, for example, the intersections of the *t*-axis of the VCO output signal u_{osc}.

The maximum value of $u_0(t)$ is the amplitude of the input signal $u_i(t)$.

The advantage of this type of phase detection is that no ripple (double frequency component) occurs in the synchronized mode. Moreover, it is easy to lock at a higher or lower frequency, as skipping one or more samples does not influence the SH output.

In Fig. 18.2 the block diagram of the PLL is given; here LPF is a low pass filter.

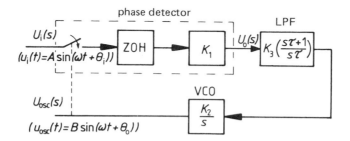

Fig. 18.2 — PLL with sample and hold device as a phase detector.

The transfer functions of the LPF and the VCO have been discussed in section 8.2.

The transfer function $H(z)$ is important for the stability investigation:

$$H(z) = Z\left[\frac{1-e^{-sT}}{s} \cdot \frac{K_L(s\tau+1)}{s^2}\right] \qquad (18.1)$$

Now follows:

$$H(z) = K_L(1-z^{-1})\mathscr{L}\left[\frac{s\tau+1}{s^3}\right] =$$

$$= K_L\left(\frac{z-1}{z}\right)\mathscr{L}\left[\frac{\tau}{s^2}+\frac{1}{s^3}\right] = K\frac{z-a}{(z-1)^2} \qquad (18.2)$$

with

$$K = K_L T^2(\tfrac{1}{2}+\tau/T) \quad \text{and} \quad a = \frac{1-T/2\tau}{1+T/2\tau} \quad \text{Please check!}$$

For small variations in the output phase (synchronized mode) K and a may be assumed constant. In Fig. 18.3 the root locus for K is sketched. If we choose $a = \frac{1}{2}$, we

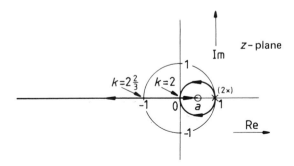

Fig. 18.3 — Root locus of a PI controlled PLL with a sample and hold device as a phase detector.

may obtain a dead beat response for certain values of K, as appears from the path of the root locus. Then the controlled system has a double pole in 0. It follows that:

$$a = \tfrac{1}{2} \rightarrow T/\tau = \tfrac{1}{3}$$

and

$$\frac{H_L(z)}{1 + H_L(z)} = \frac{K(z - a)}{z^2 - 2z + 1 + K(z - a)} \triangleq \frac{K(z - \tfrac{1}{2})}{z^2}$$

From this follows:

$$K = 2$$

The system becomes unstable for $z = -1$ and $1 + H_L(z) = 0$, so that $K = 2\tfrac{2}{3}$. Please check! So for values of K $2\tfrac{2}{3}$ the PLL is unstable.

18.3 TEMPERATURE CONTROL WITH A DIGITAL PI CONTROLLER†

A fine example of a digitally controlled process is the temperature control of Fig. 18.4.

The control is meant for analyses of acoustical properties at different temperatures.

Because a microprocessor (M6800) must be linked to the system for reasons of data logging it is obvious that we also program the control function itself. The realization is shown in Fig. 18.4.

† By courtesy of the Royal Dutch Navy Institute.

Sec. 18.3] Temperature control with a digital PI controller

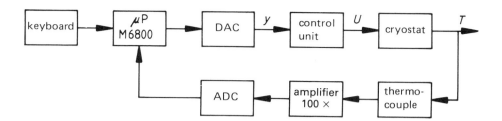

Fig. 18.4 — Digital temperature control.

The temperature is measured by the so called cryostat and is converted into a voltage by a thermocouple. The temperature ranges from −195°C to +100°C, depending on the set reference value. The controlled temperature may show an inaccuracy of ±0.1°C at most.

Because of the high accuracy of the control, we have used a 16-bit ADC and DAC. The desired temperature is set by means of thumbwheel switches.

The difference between the measured value and the set point is represented by $x(nT)$; the control signal to the process by $y(t)$.

In the cryostat there is a chamber where samples are analysed. This chamber is cooled by liquid introgen through a heat leak. To achieve the desired temperature the chamber is heated by an element (input voltage U). The steady state transfer T/U of the process appears to have a somewhat quadratic quadratic relation (applied power proportional to U^2). Because of this the system is non-linear. This non-linearity may be compensated in the microcomputer. In the temperature range considered here, the transfer function $T(s)/U(s)$ satisfies:

$$\frac{T(s)}{U(s)} = \frac{4}{80s + 1} \qquad (18.4)$$

which means a first order process with dc gain $K = 4°C/V$ and a time constant $\tau = 80$ s. The transfer fucntion of the control unit and the thermocouple is 27, and 3.5×10^{-5} V/°C respectively.

Because the thermocouple has a very low output voltage, this is first amplified by a factor of 100.

The block diagram of the control system is shown in Fig. 18.5.

Design of the analog controller
For the design of $D(z)$ we shall start from the optimal analog controller. A good controller for this purpose is a PI controller with transfer function

$$G(s) = K_R\left(1 + \frac{1}{s\tau_i}\right);$$

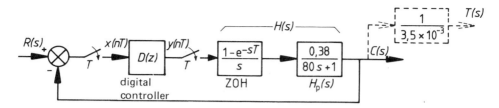

Fig. 18.5 — Block diagram of the temperature control.

In that case, the steady state error will be zero. By choosing the integration time τ_i equal to the process time constant τ, a pure integrator as a loop transfer function will develop, which degenerates into a first order system after feedback. Because the transfer function is not very stable with time, a dead-beat control is not suitable.

For the loop transfer function H_L, if $\tau_i = 80\,\text{s}$ follows:

$$H_L(s) = G(s) \cdot H_p(s) = K_R \left(1 + \frac{1}{s\tau_i}\right) \frac{0.38}{80s + 1} =$$

$$= \frac{4.75 \times 10^{-3} K_R}{s} \tag{18.5}$$

The transfer function $C(s)/R(s)$ of the controlled system is:

$$\frac{C(s)}{R(s)} = \frac{1}{s\tau' + 1} \quad \text{with} \quad \tau' = \frac{210}{K_R} \text{ seconds} \quad \text{Please check!} \tag{18.6}$$

The maximum value of K_R (and so the eventual velocity of the system) is determined by the ranges of the AD and DA converter and the control unit combined with the maximum input change (step magnitude). A value of, for example, 10 for K_R yields an effective system time constant of $\tau' \simeq 21\,\text{s}$, which is about four times faster than the original open-loop system.

Design of the digital controller

The transfer function $D(z)$ of the digital controller — as a discrete form of $G(s)$ — is:

$$D(z) = k_p \left(1 + \frac{k_i}{1 - z^{-1}}\right) \tag{18.7}$$

(see also section 14.5) with $k_p = K_R$ and $k_i = T/\tau_i$. These relations are valid only if the velocity of the system is small with regard to the sampling period, so $\tau' \ll T$. Thus:

Sec. 18.3] Temperature control with a digital PI controller

$$\tau' = \frac{210}{K_R} \ll T \quad \text{or} \quad K_R \ll \frac{210}{T} \tag{18.8}$$

As the AD converter has a fixed sampling period of $T = 4$, it is obvious that K_R must have a value for $K_R \ll 52.5$. For $K_R \geq 50$, the system no longer behaves like a first order system. Moreover, the stability of the control system may be at risk.
Assuming $T = 4$ yields, for k_i:

$$k_i = T/\tau_i = 0.05 \tag{18.9}$$

A favourable value for k_p is: $k_p \simeq 10$. For input changes that are not too large the controlled system behaves like a first order system with a time constant of $\tau' \approx 21\,\text{s}$.

Stability analysis

For the stability analysis we need the loop transfer function ($T = 4\,\text{s}$ and $k_i = 0.05$):

$$H_L(z) = D(z) \cdot H(z) \tag{18.10}$$

with

$$\left. \begin{aligned} D(z) &= K_R + \frac{0.05\,K_R}{1 - z^{-1}} = 1.05 K_R \frac{z - 0.952}{z - 1} \\ \\ \text{and} \\ \\ H(z) &= \mathscr{L}\left[\frac{1 - e^{-sT}}{s} \cdot \frac{0.38}{80s + 1}\right] = \frac{0.018}{z - 0.952} \end{aligned} \right\} \tag{18.11}$$

Here we have, for $H_L(z)$:

$$H_L(z) \approx \frac{0.02\,K_R}{z - 1} \tag{18.12}$$

The root locus of the controlled system for $0 < K_R < \infty$ is drawn in Fig. 18.6.
From the root locus appears that for $K_R = 100$ the controlled system becomes unstable. For $K_R = 50$ the controlled system has a pole at the origin, by which a dead beat control has developed. For some values of k_p ($k_i = 0.05$ and $T = 4\,\text{s}$) the step responses are shown in Fig. 18.7.

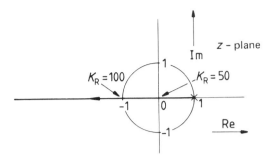

Fig. 18.6 — Root locus of the controlled system.

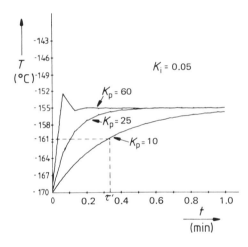

Fig. 18.7 — Step responses of the control for $k_p = 10$, 25 and 60.

18.4 SET-UP AND OPERATION OF A DIGITAL PID CONTROLLER†

This controller is for two control loops and has been constructed by means of 'standard cards' in a $\frac{1}{2} \times 19$ inch housing. The cards are able to exchange data by means of the built-in bus system; see Fig. 18.8.

There are 5 standard cards:

1. A single board computer card (SBC) on which the CPU, the PROM, and the RAM memories have been placed. The PROMs contain the programs. In the RAM we may store data that are needed only temporarily, such as results of

† LABBUS PID controller.

Sec. 18.4]   **Set-up and operation of a digital PID controller**   325

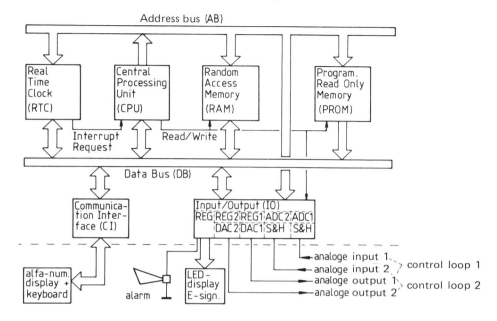

Fig. 18.8 — Block diargam of the hardware configuration of a digital PID controller for two control loops.

calculations or data that may be changed via the keyboard (e.g. control parameters).
2 A communication interface that enables data to be put in by the keyboard, or data from the system to be shown on the display.
3 A dual ADC by which two input signals (measured values) may be sampled and converted into a 10-bit word.
4 A dual DAC by which two output signals (control signals) may be formed. The registers (10 bits) realize the zero order holds.
5 A digital I/O-card of which the digital output register is used for controlling an LED indication to indicate a large error signal, by which also an acoustic signal is generated in the case of an intolerably large error signal.

The memories, the AD converters, the DA converters, and the I/O are addressed through the address bus and exchange data with the CPU through the data bus.

The operation of the controller is explained by means of the flow diagram of Fig. 18.9. At the beginning the controller will first be initiated; that is, initial conditions of control actions are read in from PROM to RAM and the registers of ADC and DAC are cleared. At this point the system comes up on to the display as 'Digital PID controller'. The control parameters of the control loops may then be put in by the keyboard and a start-up with a ramp reference signal. As soon as the system has been started up the program may be interrupted at any moment in order to execute process control, according to the left-hand part of the flow diagram.

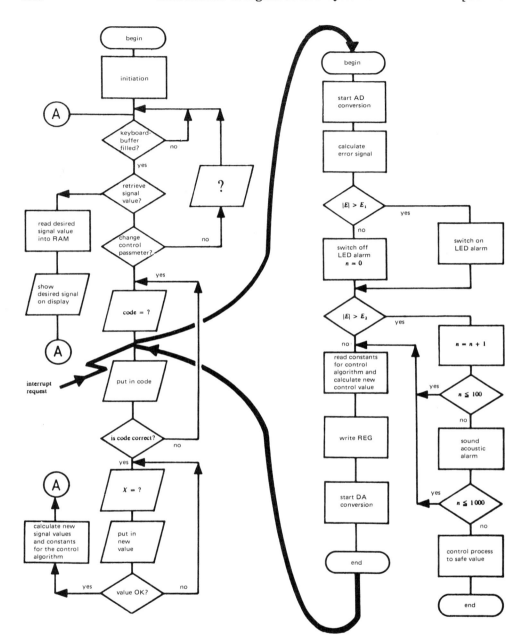

Fig. 18.9 — Flow diagram of digital PID-controller.

Interruption is done by means of the real-time clock, which issues an 'Interrupt-Request Signal' at fixed instants, set by the chosen sampling periods. The current program, according to the left-hand part of the flow diagram, will then be interrupted immediately; the present data are stored and the program on the right will be executed.

In the right-hand part of the flow diagram the process values are sampled and new control values for the process are calculated by means of the control algorithm. In this part of the program we also carry out the necessary checks by means of the error signal E. When this is larger than a given value E_1 (e.g. $\pm 5\%$ of the reference signal) we shall first get an optical alarm signal. If the error signal stays larger than the (intolerable) value E_2 for a certain time ($100T$), then we shall also get an acoustic alarm signal. During a certain time ($900T$) the operator tries to find the error. If after a time $1000T$ the error signal is still too high, then the process is put to a safe value automatically and is no longer controlled.

The reader may verify the operation of the controller by means of the flow diagram of Fig. 18.9.

The control algorithm of this controller is a discreting of the analog controller according to:

$$H(j\omega) = K_R\left(1+\frac{1}{j\omega\tau_i}\right)\left(\frac{1+j\omega\tau_d}{1+j\omega\frac{\tau_d}{a}}\right) \tag{18.13}$$

in which a is the taming factor ($1 \leqq a \leqq 20$).

At the initiation of the controller we may state wehether the position or the velocity control algorithm must be executed.

18.5 APPLICATION OF A MOVING AVERAGE FILTER

A typical application of a moving average filter is represented by the control problem sketched in Fig. 18.10.

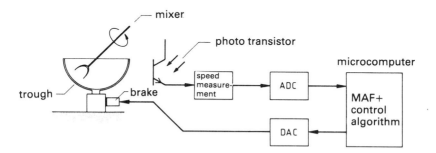

Fig. 18.10 — Speed control of a dough mixer.

Bread dough is prepared in a trough by means of a mixer. Because of the viscosity of the dough the mixer tends to accelerate. Constant speed, however, is essential for the quality of the dough. Therefore the speed is measured optically and a mechanical brake is operated by means of a microcomputer.

Analysis of the speed signal of the trough tells us that the rotation frequency of the mixer appears to be an important component. This is clearly shown by the jerky movement of the trough.

A control by which this frequency component is suppressed, must be very fast, which would take too much control energy. Hence the average r.p.m. serves as input signal, by means of a moving average filter in the microcomputers, for the control algorithm itself.

The mixer is rotating at 75 r.p.m. (constant). According to section 17.4, a fourth order moving average filter suppresses this frequency completely, if we assume a sampling period of $T = 0.2$ s. The result of the filtering is shown in Fig. 18.11.

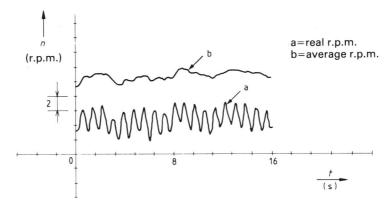

Fig. 18.11 — Effect of a moving average filter on the r.p.m. signal.

Depending on the dynamics of the process we may now determine the control algorithm further, as discussed in the previous chapter.

Answers to the problems

CHAPTER 2

1 a $RC \dfrac{dy(t)}{dt} + y(t) = RC \dfrac{dx(t)}{dt}$, $\tau = RC$

   b $F_d = F_v \rightarrow c_d \cdot \dfrac{d\{x(t) - y(t)\}}{dt} = c_v y(t)$

   $\dfrac{c_d}{c_v} \dfrac{dy(t)}{dt} + y(t) = \dfrac{c_d}{c_v} \cdot \dfrac{dx(t)}{dt}$; $\tau = \dfrac{c_d}{c_v}$

   c $\dfrac{dy(t)}{dt} = \dfrac{1}{O}(x(t) - A)$

2 $u_0(t) = 5 \cdot (3t - 2 + te^{-2t} + 2e^{-2})1(t)$

3 $y(t) = 4\{1 - \tfrac{5}{3} e^{-4t} \sin(3t + \phi)\}$
   with $\phi = \arctan \tfrac{3}{4}$

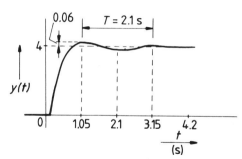

4 $\omega_0 = 4$ rad/s; $\beta = 0.4$, $\omega_g = \omega_0\sqrt{1 - \beta^2} = 3.67$ rad/s
   $K = 0.4$

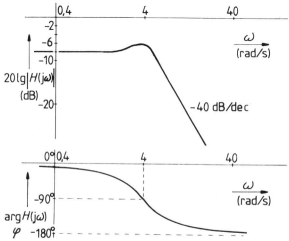

5 $y(t) = 4(1 - e^{-\frac{1}{2}(t-2)})1(t-2)$

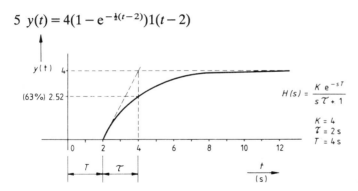

6

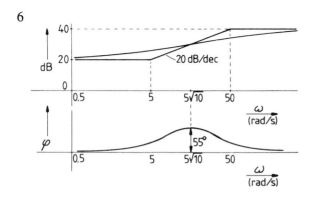

max. phase lead $\quad \phi_{max} = \arcsin \dfrac{1-\alpha}{1+\alpha}$ with $\alpha = 0.1$

$$= \arcsin \dfrac{0.9}{1.1} = 55°$$

for $\quad \omega = \dfrac{5\sqrt{10}}{5} = 15.8 \text{ rad/s}$

7

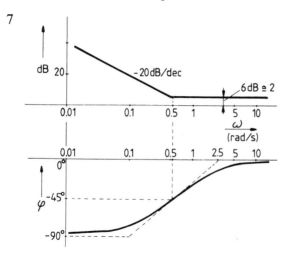

8 a $y(t) = 8t$
   b $y(t) = 1.5\{1 - e^{-t}\cos t + \frac{1}{3}e^{-t}\sin t\}$
   c $y(t) = 2\{2t - 3 + 4e^{-t} - e^{-2t}\}$
   d $y(t) = 5t^2 e^{-t}$
   e $y(t) = \frac{75}{13}\{e^{-t} - \cos 3t + \frac{2}{3}\sin 3t\}$
   f $y(t) = \frac{7}{9} - \frac{1}{3}t e^{-3t} - \frac{7}{9}e^{-3t}$

9 $X(s) = \dfrac{1 - e^{-2s} - 2s\, e^{-2s}}{2s^2}$

10 a $c(0) = c(\infty) = 0$
   b $c(t) = 2t e^{-t} + e^{-t} - e^{-2t}$
11 a $\tau = 20s$ b $T_{\text{th}}(t) = t_0 + \frac{1}{3}\{t - 20 + 20 e^{-t/20}\}$ c error = 4°C.

CHAPTER 3
1 $y(t) = 5(10t - 1 + e^{-10t})$
2

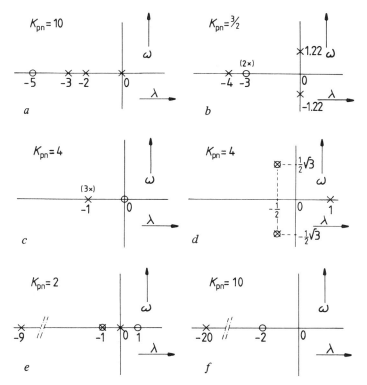

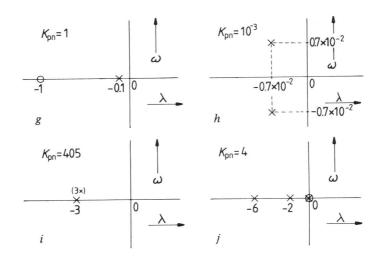

3 a $y(t) = 60(1 - e^{-t/5})$
  b $y(t) = 20 e^{-0,2t} \sin 2t$
  c $y(t) = 9 - 10 e^{-t} + 1,1 e^{-10t} - 0.01 e^{-100t}$
  d $y(t) = 48 \{\frac{1}{5} e^{-5t} + \frac{1}{3} e^{-t} \sin (3t - \phi)\}$
  with $\phi = \arctan 3/4$.

4 a $|H(j0)| = 10$ $|H(j\frac{1}{2})| = 7.07$ $|H(j\infty)| = 0$
     $\arg H(j0) = 0$ $\arg H(j\frac{1}{2}) = -45°$ $\arg H(j\infty) = -90°$

  b $|H(j0)| = 2$ $|H(j10)| = 0.02$ $|H(j\infty)| = 0$
     $\arg H(j0) = 0$ $\arg H(j0) = -168°$ $\arg H(j\infty) = -180°$

  c $|H(j0)| = 100$ $|H(j10)| = 10$ $|H(j\infty)| = 1$
     $\arg H(j0) = 0$ $\arg H(j10) = -78°$ $\arg H(j\infty) = 0$

  d $|H(j0)| = 1$ $|H(j80)| = 4$ $|H(j\infty)| = 16$
     $\arg H(j0) = 0$ $\arg H(j80) = 62°$ $\arg H(j\infty) = 0$

CHAPTER 4
1 a   unstable   b unstable   c stable   d unstable
    e stable   f unstable   g stable

3 $0 < K < 6$

4 $K > -1$

5 $t_s(5\%) = 3s$; $t_s(2\%) = 4s$; $D = 35\%$; $t_p = 1.05s$
6 To the right of $\lambda = -3$ and to the right of the lines through the origin at angles of 135° and 225°.
7 $y(t) = -1 - 0.75\,e^{-t} + 1.75\,e^{-t/5}$

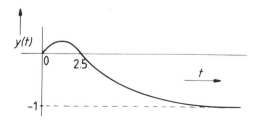

CHAPTER 5
1

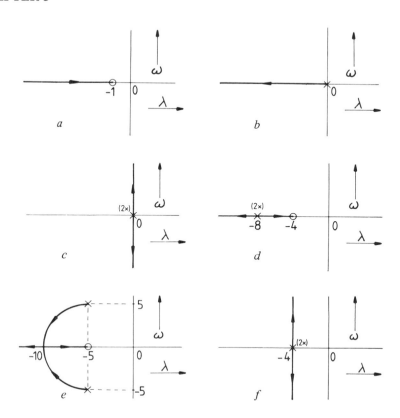

Answers to the problems

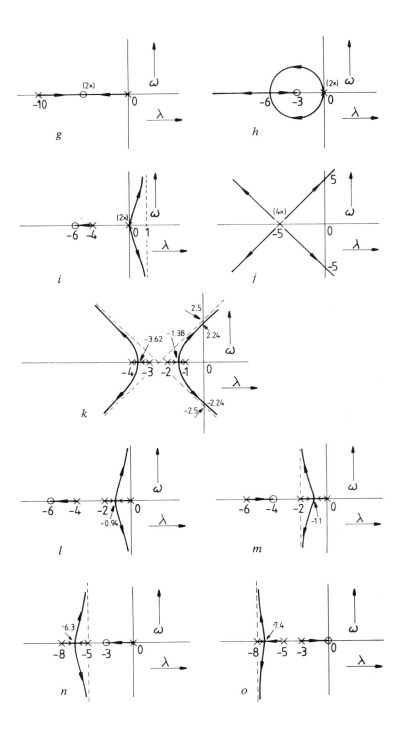

Answers to the problems

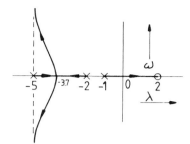

p For $K > 5$ the system becomes unstable

2 a

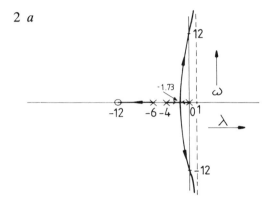

b $K_{T,osc} = 30$
$\omega_{osc} = 12$ rad/s
c $A_{max} = 40$

3 a

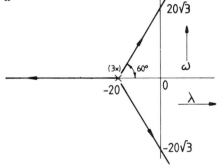

b $K_{T,osc} = 1$
$\omega_{osc} = 20\sqrt{3}$ rad/s
c $p_{1,2} = -15 \pm 5j\sqrt{3}$
$K_T = 1/16$
4 $\tau_d = 1/4$s, $a = 15$, $K = 6$
5 $K_{osc} = 16$
$\omega_{osc} = 2\sqrt{2}$ rad/s
the system is unstable for: $0 < K < 16$

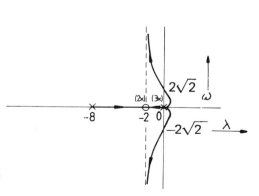

Answers to the problems

CHAPTER 6

1. $K_{osc} = 8$

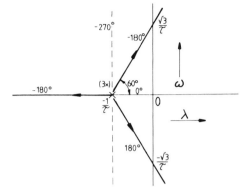

2. $K_{osc} = \dfrac{\pi}{2}, \; \omega_{osc} = \dfrac{\pi}{2}$ rad/s

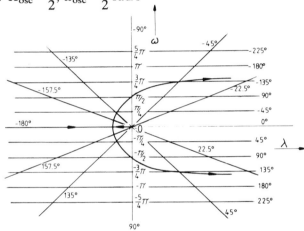

3. $K_{osc} = 0.12; \; \omega_{osc} \simeq 22.8$ rad/s

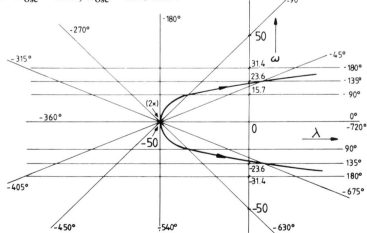

4. $K'_{osc} = 20.2; \; \omega_{osc} = 2.4$ rad/s

CHAPTER 7
1. $t_S(2\%) = 9.7$ s; $t_s(5\%) = 7.26$ s; $D = 62.4\%$ and $E_{stat} = 20\%$
2. The area to the right of $\lambda = -6$ and to the right of the lines through the origin at angles of $120°$ and $-120°$.
3. a $K = 1.9$

   b

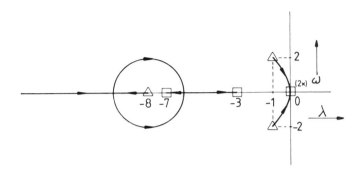

   c $\tau \leqq 25$ ms

4. a $K = 1$; $t_s(2\%) = 4$s

   b $K = 3.9$ $G(s) = 10 \dfrac{s + 7.4}{s + 74}$

   c $G(s) = 1.86 \dfrac{s + 2.9}{s + 5.4}$; $K = 2.5$

5. a $s_{1,2} = -2 \pm j4$

   c $K = 225$ $G(s) = 10 \dfrac{s + 5.43}{s + 54.3}$

   d $G(s) = 2.9 \dfrac{s + 2.6}{s + 7.5}$; $K = 132$

CHAPTER 10
2. Sampling frequency influences the dynamic behaviour; resolution affects the accuracy.
4. Velocity, accuracy, resolution.
6. $f_{max} = 124$ Hz

9 $U_0^*(o) = -\frac{18}{7}$; $U_0^*(T) = \frac{172}{49}$

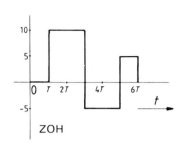
ZOH

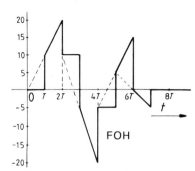
FOH

12 Replace SH by time delay $T_d = \frac{T}{2}$; system is stable for $K < \frac{\pi}{T}$.

CHAPTER 11

1 $x^*(t) = 2\delta(t - \frac{1}{2}) + 4\delta(t - 1) + 2\delta(t - 1\frac{1}{2}) + \delta(t - 2\frac{1}{2}) + 2\delta(t - 3)$
  $+ \delta(t - 3\frac{1}{2}) + \frac{1}{2}\delta(t - 4\frac{1}{2}) + \delta(t - 5) + \frac{1}{2}\delta(t - 5\frac{1}{2})$

2 $X^*(s) = \sum_{n=0}^{\infty} x(nT) \, e^{-snT} =$
  $= 2 \, e^{-\frac{1}{2}s} + 4 \, e^{-s} + 2 \, e^{-1\frac{1}{2}s} + e^{-2\frac{1}{2}s} + 2 \, e^{-3s} + e^{-3\frac{1}{2}s} + \frac{1}{2} \, e^{-4\frac{1}{2}s} + e^{-5s} +$
  $+ \frac{1}{2} \, e^{-5\frac{1}{2}s}$

3 $X(s) = \dfrac{4s}{s^2 + 10^2}$

$X^*(s) = \dfrac{1}{T} \sum_{n=-\infty}^{\infty} X(s + jn\,\omega_s)$

$= \dfrac{4}{T} \sum_{n=-\infty}^{\infty} \dfrac{s + jn\omega_s}{(s + jn\omega_s)^2 + 10^2}$

with $\omega_s = 100$.

$X^*(j\omega) = \dfrac{4}{T} \sum_{n=-\infty}^{\infty} \dfrac{j\omega + jn\omega_s}{(j\omega + jn\omega_s)^2 + 10^2}$

with $\omega_s = 100$.

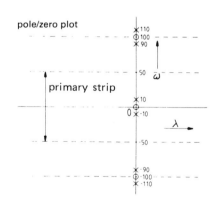
pole/zero plot
primary strip

5 a $X(z) = \dfrac{10}{z(z-1)}$

  b $X(z) = \dfrac{z}{z-2}$

  c $X(z) = \dfrac{z^4 + 2z^3 + 3z^2 + 2z + 1}{z^5}$

Answers to the problems

6 a $X(z) = \dfrac{5z^2 + 2}{z(z-1)}$

  b $X(c) = 5zT \left[\dfrac{(1-e^{-2T})(z^2 - e^{-2T})}{(z-1)^2(z-e^{-2T})^2} \right]$

  c $X(z) = \dfrac{4z[2z - 2\cos 5T + \sin 5T]}{z^2 - 2z \cos 5T + 1}$

  d $X(z) = \dfrac{z(2z - 2e^{-T} \cos 3\pi T + e^{-T} \sin 3\pi T)}{z^2 - 2z\, e^{-T} \cos 3\pi T + e^{-2T}}$

  e $X(z) = \dfrac{zT \sin \pi T\, (z^2 - 1)}{[z^2 - 2z \cos \pi T + 1]^2}$

7 $y(0) = 2;\ y(\infty) = 0$

8 a $x(nT) = -(2)^n + (4)^4$
  b $x(nT) = 1 + 2nT$
  c $x(nT) = 2\, e^{-nT} \cos \pi n T + 4 e^{-nT} \sin \pi n T$

9 a $[x(nT)] = 0, 1, 2, 3, 4, 5, 0, 0, 0, \ldots$
  b $[x(nT)] = 0, 1, 4, 14, 46, \ldots$
  c $[x(nT)] = 0, 0, 1, 0, 0, -1, 0, 0, 1, 0, 0, -1, 0, \ldots$

10 $X(z) = \dfrac{z[T + \frac{1}{2}aT^2] + \frac{1}{2}aT^2 - T}{(z-1)^2}$

11 $X(z) = \dfrac{z^2}{(z - \frac{1}{2})(z - \frac{1}{4})}$

12 $X(z) = \dfrac{zT}{(z-1)^2}$

13 a $[c(nT)] = 0, 0, \frac{1}{4}, \frac{1}{2}, \frac{11}{16}, \ldots$
   b $c(\infty) = 1$

14 a $[c(nT)] = -1, 0, 1, 2, 3, \ldots$
   b $c(\infty) = \infty$

CHAPTER 12

1 $H(z) = \dfrac{2}{z^2}$

2 $[y(nT)] = 0;\ 0.63;\ 0.86;\ 0.95;\ 0.98;\ 0.99;\ \ldots$
   $y^*(t)$ is the sampled step response $(1 - e^{-t})\ y(\infty) = 1$

Answers to the problems

3    $H(z) = \dfrac{0.37z + 0.26}{z^2 - 1.37\, z + 0.37}$

4    $[c(nT)] = 0, \tfrac{1}{8}, \tfrac{9}{32}, \tfrac{49}{128}, \ldots$
     $c(\infty) = 1/2$

5 a $C(z) = \dfrac{0.8z}{z^2 - 0.8z - 0.2}$

   b $[c(nT)] = 0;\ 0.8;\ 0.64;\ 0.67, \ldots$

6 a $H(z) = \dfrac{0.4\,K}{z - 0.6 + 0.4\,K}$

   b $[c(nT)] = 0;\ 0.4;\ 0.48;\ 0.496;\ \ldots$
       $[c(nT)] = 0;\ 0.6;\ 0.6;\ 0.6;\ \ldots$
       $[c(nT)] = 0;\ 1;\ 0.6;\ 0.76;\ 0.7;\ \ldots$

   c $c(\infty) = 1/2$
       $c(\infty) = 0.6$
       $c(\infty) = 0.714$

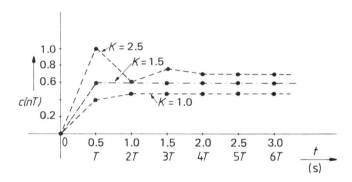

CHAPTER 13

1 a $H(z)_{\text{open loop system}} = K\dfrac{(z+2)\,(z-\tfrac{1}{2})}{z(z+\tfrac{1}{2})}$

   b $H(z)_{\text{closed loop system}} = \dfrac{K(z+2)\,(z-\tfrac{1}{2})}{z(z+\tfrac{1}{2}) + K(z+2)\,(z-\tfrac{1}{2})}$

   c

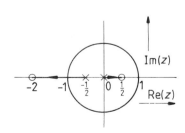

Unstable for $z = -1: \Rightarrow K_{\text{osc}} = 1/3$.

2 a $K = 1$

b
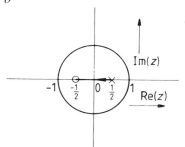

stable for every positive value of K.

3 a $H'(z) = \dfrac{0.2 K}{z(z - 0.8)}$

b
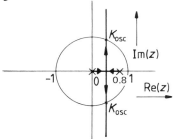

$K_{osc} \geqslant 5$

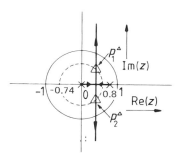

for $t_s\,(5\%) = 5$s: $K = 2.74$)

d $D = 39\%$
e $K = 0.8$; $t_s(5\%) = 1.6$ s

4 a $H(z) = \dfrac{3z - 1}{2z^2}$

b
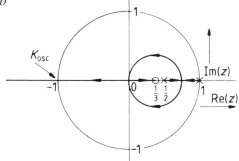

c $K \gtreqqless 9/4$

5 a $H(z) = \dfrac{1.26\,K}{z^2(z-0.37)+1.26\,K}$

b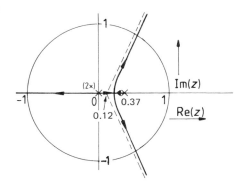

c There are three intersections with the unit circle.
For $z_1 = -1$ is $K = 1.1$.
The other poles on the unit circle are:
$z_{2,3} = 0.6 \pm \mathrm{j}0.8$ for $K = 0.66$.
So: $K_{\mathrm{osc}} = 0.66$.

CHAPTER 14
1 a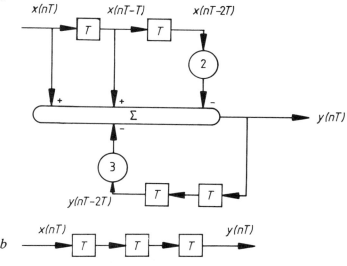

c not possible to realize: degree numerator > degree denominator.

2 $\dfrac{Y(z)}{X(z)} = \dfrac{5z^2 - 4z}{z^2 - 1}$

Answers to the problems

3 a $\quad \dfrac{Y(z)}{X(z)} = \dfrac{3(z-0.5)}{(z+0.5)(z-1)}$

b

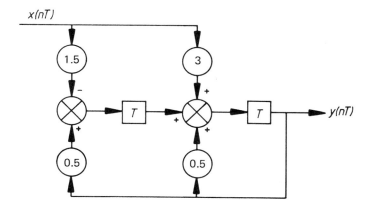

4 $D(z)_p = k_p + \dfrac{k_p \cdot k_i}{1-z^{-1}}$ with $k_p = 1$ and $k_i = \dfrac{T}{\tau_i} = 0.5$

$D(z)_p = \dfrac{1.5z - 1}{z - 1}$

$D(z)_s = \dfrac{1.5z - 1}{z}$

5 $D'(z) = k_p[(1 + k_i + k_d) - (1 + 2k_d)z^{-1} + k_d z^{-2}]$

6 $D(z) = 10k_p \dfrac{k_p\left(1 + \dfrac{\tau_d}{T} - \dfrac{\tau_d}{T}z^{-1}\right)}{1 + \dfrac{\tau_d}{10T} - \dfrac{\tau_d}{10T}z^{-1}}$

CHAPTER 15

1 a $K_{osc} = 10$, $T_{osc} = 6.3s$

$G(s) = 6\left(1 + \dfrac{1}{3.14s} + 0.8s\right)$

b $y(nT) = y(nT - T) + 5.42x(nT) - 102x(nT - T) + 48x(nT - 2T)$

2 $T_d \approx 1.3s$
$\tau \approx 7.7s$
$K \approx \approx 1$
$T = 1.5s$

$y(nT) = y(nT - T) + 2x(nT) - 2.4x(nT - T) + 0.7x(nT - 2T)$

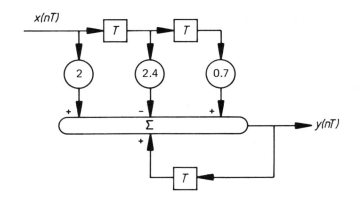

3 $T = 0.5\text{s}$ $\quad D(z)_\text{p} = \dfrac{1.05 - z^{-1}}{1 - z^{-1}}$

$$D(z)_\text{s} = 1.05 - z^{-1}$$

CHAPTER 16

1 a $\quad D(z) = \tfrac{1}{3} \dfrac{z - \tfrac{1}{2}}{z - 1}$

  b $\quad y(nT) = y(nT - T) + \tfrac{1}{3}x(nT) - \tfrac{1}{6}x(nT - T)$

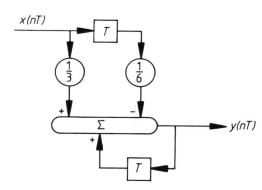

2 a

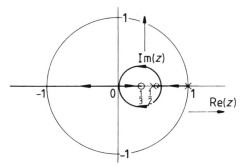

Dead-beat — the root locus passes though $z = 0$

b $K = 1.5$
c

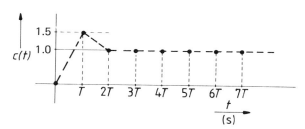

Overshoot is caused by the zero, for the 'normal' 'dead beat' response comes from $\dfrac{K}{z^2}$.

CHAPTER 17
1 $C = 0.8\mu F$
2 $y(nT) = \tfrac{1}{3}[x(nT) + x(nT - T) + x(nT - 2T)]$

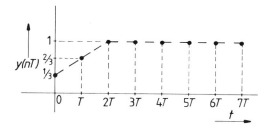

3 a $H_L(z) = \dfrac{K(z+1)}{2z^2(z-1)}$

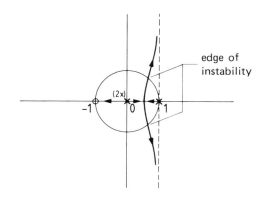

System unstable for $K \geqq 0.82$.

b $\quad H_L(z) = \dfrac{\tfrac{1}{2} K}{(z-1)(z-\tfrac{1}{2})}$

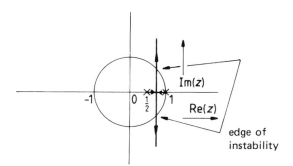

system unstable for $K \geqq 1$.

References and bibliography

1. Ogata, K.; *Modern control engineering*. Prentice-Hall, Englewood Cliffs, N.J., 1970.
2. Davidse, J.; *Ontwerpoverwegingen bij de phase-lock-loop*. Postdoctoral course Phase-Lock-Loops. Delft, 1980.
3. Underhill, M. J.; *Phase lock frequency synthesis for communications*. Symposium on Phase-Lock-Loops and Applications. Delft, 1980.
4. Verbruggen, H. B.; *Digitale regelsystemen*. Delft University Press, 1975.
5. Verbruggen, H. B., Peperstraete, J. A., & Debruyn, H. P.; *Digital controllers and digital control algorithms. Journal A*, **XVI**, no. 2, 1975.
6. Kuo, B.; *Analysis and synthesis of sampled data control systems*. Prentice-Hall, Englewood Cliffs, N.J., 1963.
7. Hendriks, S. H. & Barhorst, S. A. M.; *Digitale kryostaatregeling m.b.v. Motorola Microprocessor M6800*. K.I.M., internal report no. 1980-2. Den Helder, 1980.
8. Arends, J., Reimert, F., & Schrage, J. J.; *Inleiding tot de regeltechniek*. Nijgh & Van Ditmar, 's-Gravenhage, 1981.
9. Nauta Lemke, H. R. van, *et al.; Inleiding tot de regeltechniek*. University Press Rotterdam, 1967.
10. Nauta Lemke, H. R. van *et al.*; *Frequentieresponsies in de regeltechniek*. University Press Rotterdam, 1967.
11. Truxal, J. G.; *Automatic feedback and control system synthesis*. McGraw-Hill, New York, 1955.

Index

absolute damping line, 162
   constant, 62, 64
   in root loci, 96
   in the s-domain, 248
   in the s-plane, 58–65
A factors, 45–46
amplifier, 152
amplitude of a pulse, 179
analog–digital converter, 167–168, 180
analog filtering before sampling, 309
analog multiplexer, 180
anti-aliasing filter, 180
asymptotes, 90
automation problem, 166
'avoiding proportional and derivative kick', 277

back-up controllers, 175
back-up system, 175, 268
Bairstov, 100
bandwidth, 54, 185
'bang-bang control', 295
block diagrams, 221–239
Bode diagram, 30, 47
   of a derivative action, 126
   digital process control design in, 288–289
bumpless transfer, 268

canonical programming, 266
cascade of two blocks, 226–227
Central Processing Unit, 261
common mode rejection ratio, 181
compensation network, 261
complete digital control, 175
complex s-plane, 35
computers
   breakdown, 173
   hierarchy of, 175
   in-line applications of, 173–175
   input and output, 169
   off-line applications of, 171–172
   on-line applications of, 172–173
conjugated poles, 52
constant factor, 18, 22
constant relative damping line, 250

continuous element, 224
continuous signals, 221
control algorithms, 267–274
   examples of, 274–276
   implementation of, 276–279
control parameters, selection of, 282
controlled process, 119
controlled systems, 119–151
   proportionally, designing of, 122–126
CPU, 179
   signal processing in, 185–187
   signal reconstruction in, 187–191

D action, 24, 163, 271
   graphic determination of, 147
   in the s-domain, 139–141, 142, 144, 146
damped frequency, 30
damping ratio, 30, 162
dc gain, 24, 30, 49, 54, 58
   in the loop, 87, 88
   and root locus, 87–89
dc position servo system, 152, 153
'dead-beat control', 295
   appreciation of, 305–306
dead-beat response method, 295–299
dead time, *see* time delay
deliberate sampling, 167
derivative control action, 126–131, 270
   Bode diagram of, 126
   computer-aided design of, 141–144
   designing, 144–149
   tamed, 126
desaturation of the integral action, 277
design criteria
   in the s-domain, 120–122
   in the z-domain, 240, 247–251
designing controlled systems, 122–126
differential equation, 13–19
   linear first order, 16–17
   of second order system, 19
digital–analog converter, 167–168, 179, 180
digital filters, 308–317
digital measurements, 167
digital multiplexer, 180
digital networks, 261

Index 349

digital oscillation, 276
digital PID controller, 324–327
digital register, 252
'direct digital control', 174, 177–197
    practical rules, 282–294
direct programming, 263
discrete element, 222–223
discrete signals, 221
disturbance behaviour, 299
dominating time constant τ_1, 283
drifting of the process, 276
dual-slope ADC, 182–185
dynamic behaviour of a system, 59

exponential smoothing filter, 311

feedback
    negative, 70
    non-unit, 72
    and pole/zero representation of a system, 70–75
    in sampled data systems, 229–231
    unit, 72, 89, 231
first order hold, 190
first order system, 15
Fourier series, 204–210
frequency spectrum, 194, 207

galvanic decoupling, 180
gearbox, 152
generated frequency, 94
'gravity raiser', 152, 160–165

higher harmonics, 187

I action, 24, 271
imperfect differentiator, 18, 22
imperfect integrator, 18, 22
impulse response, 43
impulse sequence, 198–202
impure differentiator, 18
impure integrator, 18
'indirect digital control', 174
information conveying, 167
inherent sampling, 166
in-line applications, 173–175
integral control action, 131–134, 269
interfaces, 180–185
interrupt facilities, 170
intersample ripple, 302
inverse Laplace transform, 32, 45
inverse transform by division, 216–217

'keep z in hand', 216

Laplace transform, 13, 25, 26, 209
    application of, 32
    inverse, 32, 45
    of sampled data, 202–204
light-emitting diode (LED), 180
limit value theorems, 214
linearity rule, 214
logarithmic spirals, 250

low pass filter, 158, 319

magnetic force, 161
'mean time between failures', 166
mechanical load, 152
memory capacity, 170
'minimum settling time control', 295
modulation process, 200
modulator theory, 201
moving average filter, 311, 313
    application of, 327–328
multiplexers, 167
multiplexers by t, 214

negative feedback system, 70
Newton–Raphson, 100
Newton's law, 161
noise, 24, 308
noise effects, 167
non-linear conversion, 276
non-minimum-phase systems, 65–67
    and root loci, 96
non-unit feedback, 72
Nyquist's stability criterion, 54

off-line applications, 171–172
'off-line operator guide', 172
offset, 121, 131
one's complement, 184
on-line applications, 172–173
on-line operator guidance, 173
open loop of the system, 121
oscillating components, 55
oscillator, 52
output wind-up, 276
overshoot, 23, 58, 121, 289
overshoot D, 121

P action, 141, 146, 270
Padé approximation, 116
parallel blocks, 229
parallel programming, 265
peak time, 58, 63, 121
phase contribution, ϕ_d, 146
phase detector, 158
phase lag, 24, 283
phase lead, 126
phase-lines
    of a delay, 108
    of a pole, 108
    root locus constuction by means of, 109–115
    in the s-domain, 107–109
    of a zero, 107
phase lock loop, 152, 157–160
phototransistor, 161, 180
PI action, 30
PI controlled system, 132, 159
PID algorithm, practical realization of, 277
pole/zero combination, 145
    adding, 137–141
    graphical design of, 148
pole/zero plot, 37–46

350 Index

poles
  adding and varying, 136–137
  conjugated, 52
  phase-lines of, 108
  position of, 90
  presentation, 33–37
    and transfer function, 47–52
position control algorithms, 267
potentiometer, 152
prediction, 262
process automation, 166
process computer, 166
process operator, 171
processing speed, 170
proportional action, 269
proportional gain, 123
proportionally controlled system, 122–126
pulse amplitude, 179
pure differentiator, 18, 22
pure integrator, 18, 22

RC network, 310
realization diagram, 262
real-time clock, 170
reconstruction, 195
Regula Falsi, 100
relative damping line
  constant, 250
  in the s-plane, 58–65
reliability of process computers, 166
response velocity, 141
root locus, 70–105
  arrival and starting points of, on real axis, 81–84
  asymptotes of, 76–77
  computer aided calculation and, 100–102
  and dc gain, 87–89
  equation, 72, 99, 163
  in fluctuation of a time constant, 134–141
  gain, 87
  multiplication rule of, 85–87
  for negative values of K, 98–99
  and non-minimum-phase systems, 96
  parts of, on the real axis, 78–80
  phase condition of, 144–149
  and phase-lines, 109–115
  rules for, 75–98
  starting and arrival points of, 75–76
  summation rule of, 84–85
  symmetry of, 77–78
  of systems with time delay, 106–118
round-off errors, 302
Routh's stability criterion, 57, 68

sample & hold device, 167, 180
  phase-lock loop with, 318–320
  stability criteria in, 54–58
sampled data systems, 166
  block diagrams of, 221–239
  feedback in, 229–231
  first order, 251–255
  and Laplace transform, 202–204
  mathematical description of, 198–220
  second order, 245, 255–258
  transfer functions of, 223–232
  in the z-domain, 240–260
sampling frequency, 168
  choosing, 282–284
  increasing the, 309
sampling period, 167
sampling process, 185, 198
saturation, 162
s-domain, 13
  absolute damping line in, 248
  design criteria in, 120–122
  designing controlled systems in, 119–151
  phase-lines in, 107–109
  system description in, 25–28, 32–53
second order system, 18, 24, 44
  differential equation of, 19
serial programming, 264–265
servo-motor, 152
servo system, 152
settling time, 58, 120
Shannon's sampling theorem, 188, 200, 209, 283
shift rule, 214
signal filtering
  after sampling, 311–316
  before sampling, 310–311
signal frequency components, filtering out, 314
signal processing in the CPU, 185–187
signal reconstruction, 187–191
signals, processing and filtering of, 261
signalled magnitude, 184
s-plane
  complex, 35
  damping line in, 58–65
  system analysis in, 54–69
stability analysis
  in the z-domain, 240–247
stability criteria
  Nyquist's, 54
  Routh's, 57, 68
  in sample & hold device, 54–58
  in the s-plane, 54–58
stability investigation, 241
stable systems, 56, 59
steady state difference, 121
steady state error, 162, 289
subtraction point, 267
successive approximation (ADC), 181–182
superposition, 16
'supervisory control', 174
systems with time delay, 67–68, 106–118

tachogenerator, 152
taming factor, 126, 163
τd locus, 143
Taylor's expansion, 189
t-domain, 13–20
  pole-zero plot and responses in, 37–46
temperature control, 320–324
  digital, 321
thermocouple, 321

Index

time constant (τ), 14
  dominating τ_1, 283
  fluctuation of, 134–140
time delay, 19, 23
  approximation methods of, 115–117
    accuracy of, 116
  systems with, 67–68, 106–118
time-optimal control systems, 295–307
time response, 38
time sharing, 167
time shift, 262
transducer, 167
transfer function
  and pole presentation, 47–52
  of sampled data systems, 223–232
transfer ratio, modules of, 23
two's complement, 184

undamped frequency, 30
unit circle, 244
unit feedback, 72, 89, 231
unit step, 42
unstable systems, 56

velocity control algorithms, 267
voltage controlled oscillator (VCO), 318

ω-domain, 20–25, 47
word length, 170

z-domain, 13
  design criteria in, 240, 247–251
  sampled data systems in, 240–260
  stability analysis in, 240–247
  system description in, 28–29
zero order hold, 190
zeros
  adding and varying, 134
  phase-lines of, 107
  position of, 90
  presentation, 33–37
  and transfer function, 47–52
Ziegler and Nichols adjustment rules, 284, 287
z-plane, 242, 246
z-transform, 29, 210–213
  inverse, 215–217
  properties of, 213–215
  table of, 213–215

ELLIS HORWOOD SERIES IN
ELECTRICAL AND ELECTRONIC ENGINEERING

Series Editor: PETER BRANDON, Emeritus Professor of Electrical and Electronic Engineering, University of Cambridge

APPLIED CIRCUIT THEORY: Matrix and Computer Methods
P. R. ADBY, University of London King's College
NOISE IN ELECTRONIC DEVICES AND SYSTEMS
M. J. BUCKINGHAM, Royal Aircraft Establishment, Farnborough, Hampshire
MOBILE CONTROL OF DISTRIBUTED PARAMETER SYSTEMS
A. G. BUTKOVSKIY, Academy of Sciences of the USSR, Moscow
DIGITAL AND MICROPROCESSOR ENGINEERING
S. J. CAHILL, Ulster Polytechnic
ADVANCED MICROWAVE ENGINEERING
R. CHATTERJEE, formerly Indian Institute of Science, Bangalore
POWER LASERS
J.-F. ELOY, National School of Physics, Grenoble, France, and Research Staff Physicist, French Atomic Agency
INTRODUCTION TO MICROCOMPUTER ENGINEERING
D. A. FRASER, Chelsea College, University of London
NON-LINEAR AND PARAMETRIC CIRCUITS
F. KOURIL and K. VRBA, Technical University of Brno, Czechoslovakia
WAVE DIGITAL FILTERS
S. S. LAWSON, City University, London
ELEMENTARY ELECTRIC POWER AND MACHINES
P. G. McLAREN, University Engineering Department, Cambridge
MICROWAVE PHASE MODULATORS
T. MORAWSKI and J. MODELSKI, Warsaw Technical University
OPTICAL AND OPTOELECTRONIC DEVICES AND SYSTEMS
DOUGLAS A. ROSS, University of Colorado at Denver
PRINCIPLES OF COMPUTER COMMUNICATION NETWORK DESIGN
J. SEIDLER, Institute of Fundamentals of Informatics, Polish Academy of Sciences
ELECTRICAL INTERFERENCE AND PROTECTION
E. THORNTON, formerly Ministry of Defence, Aldermaston

ELECTRONIC AND COMMUNICATION ENGINEERING
TELECOMMUNICATIONS TECHNOLOGY
R. L. BREWSTER, University of Aston in Birmingham
DIFFRACTION THEORY AND ANTENNAS
R. H. CLARKE and JOHN BROWN, Imperial College of Science and Technology, University of London
PARALLEL ARRAY PROCESSING
P. G. DUCKSBURY, Systems Designers Scientific, Gloucester
SATELLITE BROADCASTING SYSTEMS: Planning and Design
J. N. SLATER, Independent Broadcasting Authority, Winchester, and L. A. TRINOGGA, Leeds Polytechnic
CABLE TELEVISION TECHNOLOGY
J. N. SLATER, Independent Broadcasting Authority, Winchester
SIGNAL CODING AND PROCESSING: An Introduction Based on Video Systems
J. G. WADE, Plymouth Polytechnic

«ML

DESIGNING ANALOG AND DIGITAL CONTROL SYSTEMS